Tales of an Old Ocean

Tales of an Old Ocean

Tjeerd van Andel

W · W · NORTON & COMPANY, INC.
New York

 Published simultaneously in Canada by George J. McLeod Limited, Toronto. Printed in the United States of America.

This book was published originally as a part of THE PORTABLE STANFORD, a series of books published by the Stanford Alumni Association, Stanford, California. This edition first published by W. W. Norton & Company, Inc. in 1978 by arrangement with the Stanford Alumni Association.

Library of Congress Cataloging in Publication Data

Andel, Tjeerd Hendrik van.
Tales of an old ocean.

(The Portable Stanford series)
Bibliography: p.
I. Oceanography. I. Title.
GC11.2.A46 1978 551.4′6 77–17958
ISBN 0–393–03213–2
ISBN 0–393–00883–5 pbk.

1 2 3 4 5 6 7 8 9 0

REDITS

OVER & AGE 158	Klee, Paul. *Battle scene from the comic operatic fantasy "The Seafarer."* 1923. Watercolor and oil. 15" x 20¼". ©by S.P.A.D.E.M., Paris, 1977. Collection, Trix Dürst-Haass, Basel. Photo: Colorphoto Hinz, Allschwil-Basel.
AGE xii	Klee, Paul. *Around the Fish.* 1926. Oil on canvas. 18⅜" x 25⅛". Collection, The Museum of Modern Art, New York. Abby Aldrich Rockefeller Fund.
AGE 5	Turner, Joseph Mallord William. *Rockets and Blue Lights (Close at Hand) to Warn Steamboats of Shoal Water.* 1840. Oil on canvas. 36-1/16" x 48⅛". Sterling and Francine Clark Art Institute, Williamstown, Massachusetts.
AGE 12	Vedder, Elihu. *Memory.* 1870. Oil on mahogany panel. 21" x 16". Los Angeles County Museum of Art. Mr. and Mrs. William Preston Harrison Collection.
AGES 24 25	Durán, Diego. *The Envoys of Moctezuma Observing the Ships of Grijalba.* 16th century. Illustration to the manuscript *Historia Destas Yndias y Yslas de tierra firme del Mar oceano.* Biblioteca Nacional, Madrid.
AGE 30	Whistler, James A. McNeill. *Nocturne in Black and Gold: Entrance to Southampton Water.* 1872 or 1874. Oil on canvas. 20" x 30". Courtesy of The Art Institute of Chicago. The Stickney Collection.
AGE 37	Bradford, William. *In Polar Seas.* 1882. Oil on canvas. 28" x 48". Private collection.
AGE 54	Etruscan fishing. Detail of Etruscan mural painting from a tomb at Tarquinia. ca. 520-510 B.C. Courtesy of Walther Drayer, Zurich.
AGE 68	Roman mosaic. Provincial Archaeological Museum, Tarragona, Spain.
AGE 72	Unknown artist. *Meditation by the Sea.* 1850-60. Oil on canvas. 13½" x 19½". Courtesy of The Museum of Fine Arts, Boston. M. and M. Karolik Collection.
AGE 74	Brueghel, Pieter. *The Hunters in the Snow.* Detail. 1565. Oil on panel. 46" x 63¾". Kunsthistorisches Museum, Vienna.
AGE 79	Whittredge, Worthington. *Third Beach, Newport, R.I.* 1865. Oil on canvas. 18" x 30". Private collection. Kennedy Galleries, Inc., New York.
AGE 91	Benton, Thomas Hart. *Cradling Wheat.* 1938. Tempera-oil on board. 31" x 38". Courtesy of the St. Louis Art Museum.
AGE 98	Brueghel, Pieter. *The Numbering at Bethlehem.* Detail. 1566. Oil on panel. 45-11/16" x 64¾". Musées Royaux des Beaux-Arts, Brussels.
AGE 100	Moran, Thomas. *Seascape.* 1906. Oil on canvas. 30" x 40". The Brooklyn Museum. Gift of the Executors of the Estate of Colonel Michael Friedsam.
AGE 112	*Departure from Venice.* Detail. Illustration to the manuscript *The Travels of Marco Polo (Livre du Grand Caam).* ca. 1400. Bodleian Library, Oxford.
AGE 116	Fortuny, Mariano (1838-1874). *African Beach.* Detail. Watercolor. Museo de Arte Moderno, Barcelona, Spain.
AGE 118	Anonymous English artist. *A Sea Flight.* Illustration to manuscript (Marlay Add. MS.I, f. 86r). ca. 1270. Courtesy of the Fitzwilliam Museum, Cambridge.
AGE 136	Klee, Paul. *Mild Peril at Sea.* 1928. Pen and ink. 17-15/16" x 12-13/16". Paul Klee Foundation, Museum of Fine Arts, Berne. © by COSMOPRESS, Genève & S.P.A.D.E.M., Paris, 1977.
AGE 142	U.S. research submersible *Alvin.* Photograph by Emory Kristof. Copyright National Geographic Society.
AGE 154	Blakelock, Ralph Albert. *The Sun, Serene, Sinks into the Slumberous Sea.* Oil on canvas. 16" x 24". Museum of Fine Arts, Springfield, Massachusetts. Bequest of Horace P. Wright.
AGE 160	Copley, John Singleton. *The Return of Neptune.* ca. 1754. Oil on canvas. 27½" x 44½". The Metropolitan Museum of Art. Gift of Mrs. Orme Wilson, 1959, in memory of her parents, Mr. and Mrs. J. Nelson Borland.
AGE 164	Natasa Sotiropoulos. *Sketch of the author.* Pen and ink. Courtesy of Tjeerd van Andel.

gures for this book have been drawn by Perfecto Mari of the School of Earth Sciences at Stanford, except r Figures 6, 6a, and 11. Cover graphics by Mark Olson.

For Marjorie, Christopher, and Jeffrey
who bore with patience the ebb and flow
of my creativity.

FOREWORD

"Piglet, I have decided something."
"What have you decided, Pooh?"
"I have decided to catch a Heffalump."
A.A. Milne, *Winnie-the-Pooh*, 1926

A FEW TIMES in the last 20 years I have toyed with the idea of writing a book, always on a subject central to my scientific expertise. When I saw colleagues struggle endlessly with such a task, however, and observed at geological conventions the massive displays of books I would never find the time to read, I always managed to dissuade myself. A quotation from an author whose name I have never known keeps coming to my mind, "And after the party they all went home and wrote books." Perhaps it would have been better if they had gone to bed. Now I find myself writing one, and doing so in a period of my life that is filled with change and urgent occupations. An explanation would seem to be in order.

Quite a few things have contributed to my decision to give in to some expert arm-twisting. The most important of these is my growing appreciation of the value of adult education. The custom that only once, always quite early in life, one prepares oneself by methodical learning for a career, then dedicates the remainder of one's years to more productive enterprises, no longer seems very appropriate, and yet society provides few means for continuing education and reorientation. As I grow older I am becoming more and more aware of the gradual decline of my skills and knowledge compared to those of younger, more recently educated colleagues, an inevitable consequence of the rapid change of science and society.

More than that, however, I begin to see that the end of youth does not bring the end of interest in and opportunity for new experiences and thoughtfully selective learning, and that society would be well served if there were more opportunities for people to revitalize their interests and bring their wisdom to bear on new subjects. A good case in point is the increasing mean age of university faculties that results from stable or declining enrollments. The prospect of a university staffed almost entirely by professors over 50 years old is not an inspiring one; decades of teaching and scholarly research in a sharply defined field of specialization have dimmed the freshness of many, although certainly not all. An opportunity to enter solidly into a new field with all the curiosity and challenges that it would entail might cause many of them to contribute anew from sources of old wisdom and fresh enthusiasm. The practical means to do so, however, are usually not there, and creating them would seem to be an urgent task. It is obviously difficult, if not impossible, to interrupt life in mid-career and return productively to a school system designed for the young. Yet so much talent, potential productivity, and happiness are lost in all walks of life by people forced to adhere to a career chosen at an age when they were least able to make a wise choice.

Another persuasion is the possibility to write, for once, about a wide and varied field in something approaching its full extent and to include aspects about which I cannot call myself a credentialed expert. It is the strength and the curse of most of us that we must be specialists and are compelled, by modesty and timidity, to refrain from expressing the thoughts and impressions we have collected over the years about contiguous subjects. Some live happily within a confinement that provides security and the opportunity for a fine honing of skills. Others find it more natural to roam widely. Being, as it were, only by accident an oceanographer, I count myself in the last group.

After the usual childhood problems of choosing between being a sea captain and an architect, I decided (living in Indonesia at the time) to become an Indonesian archeologist. I held onto this dream until I had received my bachelor's degree, but after World War II it was obvious that the demand for Dutch-born Indonesian archeologists would be very small, and I resolved to detour into geology to study the influence of the environment on early man and his effect on it. The necessity of graduate support led me into sedimentology as a possible pathway, but it turned out to be a pathway into petroleum geology instead. Because of the illness of a former professor, who had been hired by my company to do a marine geological study, I was made his substitute and faced the challenges of the sea for the first time in my life. Subsequently I published a well-received paper on our findings, was hired by the Scripps Institution of Oceanography, and, by this circuitous route, found my permanent career. All this has worked out rather well because I discovered a love for the life at sea and an abiding interest in a science so involved with human affairs. Peripheral to my work in marine geology and geophysics, I have been involved in economic aspects of marine resources, in the politics of the ocean, and in Navy issues. I have recently become intrigued with the impact of the sea on human values and with the influence of scientists on society and society on scientists.

This book deals with many facets of ocean science, although it reflects, of course, my geological knowledge and leanings. Its limitations are those of my personal experience. The choice of subjects may appear at first to be somewhat random. I would like to think that, apart from these personal interests and constraints, the connecting thread is science in its manifold relations to human life, the pure joy of knowing, the economics of resources, and the art of living together in a complex world.

Recently I was given the opportunity to explore segments of ocean science with a group of Stanford alumni on the 1976 Alaska Oceans College. The experience was exhilarating and the participants of the College have had a great deal to do with my decision to put some of the material on paper. In particular, I am grateful to Betty and Marron Kendrick and to Jean and Paul Hanna who convinced me that I had something worthwhile to say.

Owing to the rapid progress of ocean science in the last decade, the literature is very technical, and syntheses and readable reviews are in short supply. It would seem both pedantic and of little use to document my story in detail with references to this esoteric body of information. Therefore, with apologies to the colleagues whose work I present and whose facts and thoughts I have sometimes appropriated, I have restricted the Reader's Guide to material helpful to those who want to read further, and have furnished no notes. The views and concepts contained in the following chapters are only partly mine; where they are not, any distortions are my responsibility.

With a scope as large as this book I evidently owe much to many colleagues and students. I can thank them only collectively for the inspiration they have provided over many years. My editor, Cynthia Fry Gunn, has not only greatly improved the style and clarity of the text but is also responsible for the choice of the art illustrations. I owe her much for her fine collaboration. I thank also Perfecto Mari of the School of Earth Sciences at Stanford University for the drawings that clarify the text.

Stanford, California *Tjeerd H. van Andel*

TABLE OF CONTENTS

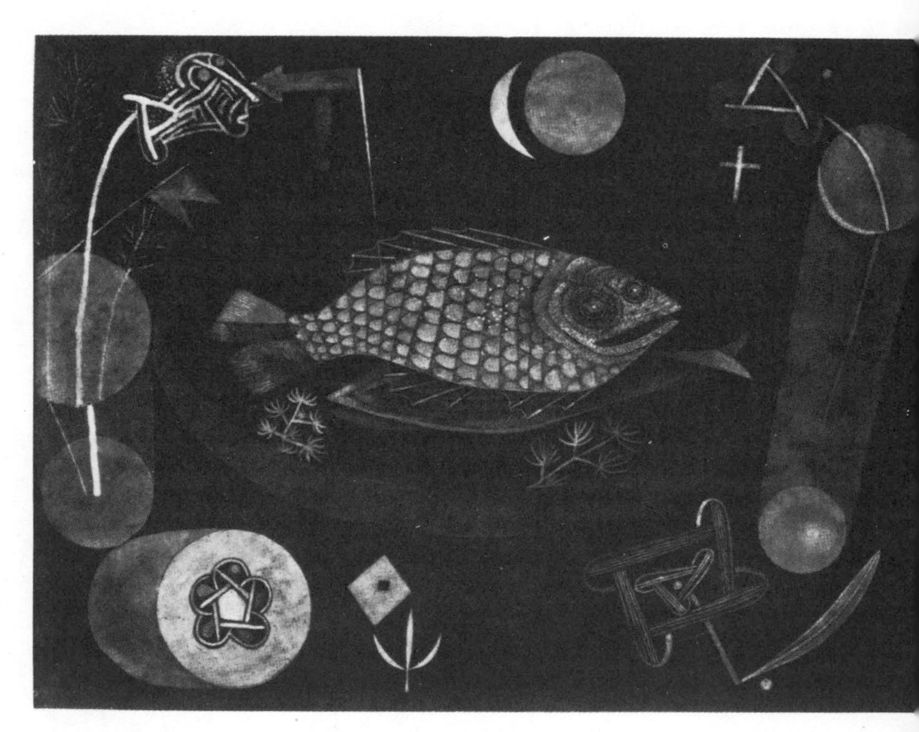

Around the Fish by Paul Klee. Collection, The Museum of Modern Art, New York. Abby Aldrich Rockefeller Fu

INTRODUCTION

QUEST FOR THE MYSTERIES OF THE SEA

> *The great Master of Philosophy drowned himself because he could not apprehend the Cause of Tydes; but his Example cannot be so prevalent with all, as to put a Period to other Mens Inquiries into the Subject.*
>
> Richard Bolland, *A Draught of the Streights of Gibraltar,* 1675

IT IS DIFFICULT to think of a science that touches simultaneously on more different aspects of human affairs than does oceanography. The sea contains many and varied resources essential to man's well-being: salt, fish, chemicals, and metals. Petroleum is being produced from offshore reservoirs and is believed, rightly or wrongly, to hold the key to the solution of our country's energy crisis. Because ore resources on land are becoming exhausted, exploration has begun for nickel, copper, cobalt, and other metals in the sediments of the ocean floor. Someday, perhaps, hard-rock mining in the deep ocean will become possible. For more than a decade, England has depended on the seabed to provide most of its sand and gravel needs, and Texas obtains much of the lime it uses for cement from old oyster beds in nearshore waters.

The sea has also long served as a convenient and economic means of transportation and as a stage for the exercise of military power, a strategic role undiminished by air power. More exotic uses of the sea are in the offing—offshore nuclear power plants, energy from tides and waves, perhaps floating cities. Many planners believe that by the year 2000 the offshore industrial installations of the United States will represent an investment of more than one trillion dollars, which raises some interesting questions about the protection of this investment against natural and man-made disasters.

The seashore has always been a popular place for recreation and has become, increasingly, a focus of human habitation with all the associated problems—use conflicts, environmental decay, overcrowding, and so forth. Finally, the sea has strongly attracted the human spirit—to seek adventure, to satisfy a basic curiosity about nature, to tap creative resources in the arts and letters.

It is curious that our deep involvement with and dependence upon the sea are, from the perspective of millions of years of human history, comparatively recent phenomena. The archeological record shows that for most of prehistoric time man was strongly land-oriented. Then, a mere 10,000 years ago, shore-based cultures began to appear in abundance, flourished, and rapidly developed into the great ancient maritime civilizations—the Phoenicians, Carthaginians, and Greeks, to name a few. This drastic change in outlook may be an artifact of the preservation of prehistoric remains at the unstable and shifting margin of the sea, but if it is not it is astonishing and not understood. Yet this turn to the sea has certainly proved to be very fruitful for the further development of man and civilization. Communities based on marine resources, sea trade, or naval power have dominated the advance of man for the last 5,000 years.

A Brief History of Oceanography

The basic understanding of the sea which should be associated with a maritime civilization has been even slower in developing than man's discovery of the ocean as his legitimate habitat. Even the blossoming of the sciences that began in the 17th century led but sporadically to an interest in problems of the sea, and that only for practical reasons such as the prediction of tides or the invention of better means for navigation (a problem that now and then still plagues modern oceanographers). In the early 19th century, when numerous physicists, chemists, biologists, and geologists were addressing fundamental problems of their fields, ocean scientists were few in number and usually only temporarily attracted by the sea. Margaret Deacon, a British historian of science, has examined this curious lack of vitality in such an important field of endeavor and concluded that it was due to the vastness of the subject and hence the large number of people and the great amounts of money that were required.

The early oceanographers themselves, by and large, would have agreed with this view. In the late 18th century, for example, James Rennell, an early student of the subject, remarked that ocean science would bloom only if governments would support it. Governments, however, tend to hold narrow and traditional views of how to spend the taxpayer's money and did not follow this suggestion until much later. In Great Britain, government support became significant in the late 19th century, but in the United States during the early 1800s the Constitution was interpreted as specifically forbidding the support of science. (In 1807 the House of Representatives told a hapless astronomer: "It is [our] opinion that application to Congress for pecuniary encouragement of important discoveries . . . cannot be complied with, as the Constitution . . . appears to have limited the powers of Congress to granting patents only.") Consquently commercial and naval interests had to be invoked in order to gain Congressional funding for early ventures such as the coastal surveys on the East Coast beginning in 1832 or the study of tides and currents by America's first great oceanographer, Matthew Fontaine Maury, in the second quarter of the 19th century. By contrast, inland explorations were common and enjoyed considerable moral and even some financial support, even though their claims of utility were often thin and barely disguised the burning interest of the explorers in nature for its own sake.

Gradually, however, the 19th century began to produce fundamental changes in man's attitude toward the sea. Previously, even the most celebrated of the 17th-century Dutch seascape painters viewed the sea merely as a backdrop against which to display ships and naval battles. The elaborate diaries of the voyages of the great world explorers seldom mentioned the sea at all except to complain of the hardships the elements inflicted on them. At least on paper, the explorers and the novelists whose imagination they captured tended to think of the sea primarily as a convenient means of getting from here to there. The real novel of the sea, poetry of the sea, and paintings evoking the moods of the sea itself came in the 19th century, with J.M.W. Turner the forerunner and Herman Melville and Joseph Conrad the bloom.

The sciences were late in joining this trend toward a less mercenary and more spiritual interest in the sea, but in 1872 the British frigate H.M.S. *Challenger* set sail on what was to be the first truly modern oceanographic expedition. In three and a half years, traversing all of the world's oceans, the expedition gathered a volume of knowledge on seawater, currents, life, and the ocean floor that still forms an essential part of our oceanographic knowledge. The horizons opened by the *Challenger* inspired a considerable amount of activity in Europe which eventually led to a flowering of the ocean sciences in the first half of the 20th century that had little to do with expectations for practical applications (except perhaps in the field of fisheries). This was the era of the globe-girdling cruises of the ships with the romantic names like *Galathea, Meteor, Albatross*—cruises that took years to complete and decades to record in massive tomes. My thesis supervisor, Professor Philip Kuenen, who participated in one of these expeditions, the voyage of the *Snellius* in Indonesian waters in the 1920s, once remarked to me that the prime requirement for participation was an ability to get along with others in close quarters for a long time, not scientific acumen. He did not recommend the experience.

Many countries participated in this phase of global exploration of the sea which lasted until World War II—Britain, Denmark, Germany, Sweden, and the United States among others—but what was lacking was a stable home base where the work could continue more or less permanently and the aggregate experience be accumulated. A real oceanographic institution had been established in Monaco around 1900 under patronage of Prince Albert I, but in most countries the level of support and the locale of the activity changed both frequently and unpredictably, with scientists migrating in and out of ocean science repeatedly. Little attention was given to education in oceanography or the creation of a stable manpower pool.

The United States was not a vigorous participant in the global voyages, but distinguished itself in the years between 1900 and World War II by establishing two permanent oceanographic institutions which, at first purely through private support, came to dominate the scene in the U.S. and abroad. The Scripps Institution of Oceanography in La Jolla, California, was founded in 1903 with support from the Scripps family. It became part of the University of California in 1912, but remained very nearly independent for many years. On the East Coast, the Woods Hole Oceanographic Institution was established in 1930 on Cape Cod with funds from the Rockefeller Foundation. Oceanography in the U.S. was on its way, but for many years its life-style remained modest by present standards, being meagerly funded by a strange assortment of supporters, staying close to home in its explorations, and surviving on the dedication of a small pioneer staff. These early oceanographers must have been driven by strong convictions and an intense curiosity about the sea, because few other rewards could be had. During that same period, efforts by government agencies, mainly the Navy, the Coast and Geodetic Survey, and fisheries agencies, were confined to practical and generally modest objectives.

The status of ocean science changed dramatically during World War II when the United States once again, as it had done in World War I, discovered the military uses of science in general and oceanography in particular. Antisubmarine and amphibious operations required a great deal of

ockets and Blue Lights (Close at Hand) to Warn Steamboats of Shoal Water by Joseph Mallord William Turner.

research and weapons development, and the ocean science establishment ballooned. An entire generation of oceanographers, including most of the present great men, earned their credentials in the 1940s, and nearly all of them drifted in, having had their formal training in other more basic fields such as physics, geology, or engineering. Higher degrees in oceanography itself had been occasionally obtained at Scripps since 1919, but the appearance of large numbers of young PhDs in the field did not come until the late 1950s.

Subsequent events proved that people like James Rennell, who saw salvation for oceanography only in government support, were quite correct. During the 1950s and early '60s there always seemed to be enough money, first from the Office of Naval Research and later from the National Science Foundaton. Several new oceanographic institutions grew up overnight, and the number of scientists and their professional caliber and sophistication increased exponentially. Fancy equipment multiplied. In 1951 on my first oceanographic cruise I was able to repair every piece of equipment myself with a screwdriver and a pair of pliers; today an entire team of technicians is needed to take care of maintenance and even operation. The oceanographic fleet grew by leaps and bounds. Going to sea, instead of being a once-in-a-lifetime adventure, became a yearly and sometimes monthly routine. Needless to say, the level of expenditures to support all this grew proportionally. Most of the activity was directed toward a fundamental understanding of the processes operating in the ocean, the shallow and deep circulation of seawater, the dynamics of biological communities, and the origin of the ocean basins and their sediments. Although practical aspects such as fisheries biology or the forecasting of wave conditions were not neglected, the applied aspects of the science were not foremost in the minds of most practitioners, and the government rarely, if ever, inquired whether our endeavors were of any use.

It is obvious that such a happy state of affairs could not last indefinitely. The cost of ocean exploration had risen considerably and some of the plans, such as the one to drill a deep hole in the ocean floor to take a sample of the rocks below the crust of the earth, had attracted a lot of public attention, not all of it favorable. This project—called Mohole because it was designed to drill below an important level of change in the crust, the Mohorovicic discontinuity (or Moho)—was born as a Sunday morning dream of a few scientists and acquired a reality well beyond anyone's expectations. During the period of planning, design, test drilling, and final construction of the drilling rig in the late 1950s and early '60s, Mohole was constantly in the news and the cost estimates grew from a few million to well over $100 million before the project was finally abandoned in 1965 as too costly. During the unhappy life of Mohole, scientists learned that such expensive dreams are tempting to entrepreneurs and easily become political issues.

By now, the growing body of knowledge had revealed the potential of hitherto unknown marine resources, such as gold and diamonds on continental shelves and copper and nickel in the manganese nodules of the deep sea. On occasion these and other potential contributions to the prosperity of the nation were strongly promoted and quite often oversold, and the image of the ocean as a new frontier of unlimited opportunity caught on quickly. There suddenly was much talk about the "national marine effort," and federal agencies and industry discovered abruptly that they had, after all, considerable stakes in this land of opportunity. To a large extent, these hopes were founded on the belief that the federal treasury could be persuaded to disburse funds proportional to the amount of activity that was being contemplated, just as had been the case in previous years on a very much more modest scale. Inevitably, the politics of the ocean was born and coordination of the multiple federal, academic, and industrial claims became necessary. In 1966 the National Council on Marine Resources and Engineering Development was established, an organization as ponderous as its name. The history of this council (which survived until 1971) and the various turmoils and donnybrooks surrounding its birth have been described in fascinating detail by Edward Wenk in *The Politics of the Ocean*. The book sketches the strange world in which the academic oceanographers suddenly found themselves enmeshed. Gone were the days when we controlled our own destination and, by knowing all the players, could keep the game simple and mostly honest. Suddenly many senior scientists, without much training or warning, found themselves engaged primarily in politics, fund raising, management, and endless, endless committees, panels, and workshops—much to the detriment of their research.

Toward the end of the 1960s, under the impact of the war in Vietnam and the growing awareness of widespread social ills that demanded costly cures, there was increasing pressure to restate the nation's research and development programs in terms of "national needs"—rather vaguely expressed expectations of material or social benefits. Although these national needs have never been incorporated in a specific national science policy, they have had a large influence on research directions because of their emphasis on short-term practical applications. Moreover, a sudden national awareness of the deterioration of the environment, with possibly serious effects on the ocean and its resources, opened up a whole new field of applied ocean science and, along with it, vast opportunities for conflicts, power plays, and the diversion of energy from basic research.

One beneficial side effect of the research activity and the publicity suddenly surrounding it has been the growing interest of the public in the ocean, its life, its processes, and the beauty and strangeness of this unknown world. Television programs and books on the sea are common now, and a Jacques Cousteau film attracts audiences as large as the products of more conventional entertainment. Sometimes, the output of science can have great value as a leisure-time occupation of the highest order. On this basis, public support could probably be defended, but, alas, the argument has never worked.

The State of Ocean Science Today

Thus the stage has become very crowded and the stakes for the players are high. Even to the eye of the insider the situation appears everchanging and often confused, putting a high premium on a reliable pipeline of information to Washington. A few trends in ocean science today deserve a closer look.

The first of these trends is the increased emphasis on a rapid, preferably immediate return for research investments. As a result, applied research and development have become much more attractive words than basic research, even in oceanography where so little is known as yet. It is interesting that, while barely ten years ago the federal agencies fell all over themselves to show how basic their research programs were, now the academic scientists go out of their way, not always totally honestly, to demonstrate how *soon* their results are likely to benefit mankind. In fact, under the new ground rules of many federal sponsors, they have to. There is also much greater participation in research and development by the federal government itself than there was a decade ago. Concern with the oceanic environment and interest in marine mineral resources have thus brought some new players, such as the United States Geological Survey, who compete in part with the academic community for funds and fame.

Another change of great importance is the strong shift to large, multi-institutional research projects. Oceanography has always been, of necessity, a team effort—its interdisciplinary nature and the use of costly ships make that mandatory. In the last ten years, however, the trend to massive cooperative enterprises has been greatly accelerated, not the least because such projects have much public glamour and are, because of their size, easier to manage for the funding agency, though not for the academic scientists that are involved. This shift is changing the whole outlook of ocean science. Large projects consume a great deal of time for planning, fund raising, and management—time that cannot be spent on research. Their very large budgets lead inevitably to a more careless use of money. Difficult communications within a large, far-flung team of scientists reduce effectiveness. And finally, large sets of expensive equipment and supporting technical personnel, once established, demand to be used.

As a result, once a team is on the "big project bandwagon" it is very difficult to get off without wholesale liquidation of entire staffs or the mothballing of valuable equipment. Thus scientists barely back from the sea on one project will instantly start planning another. In the crunch, the years of effort needed to digest and analyze the data are cut back sharply or even left entirely to outsiders. A meteorologist, in whose field this syndrome is also common, calls the phenomenon *Flucht nach Vorne,* "the escape into the future." No remedy has been discovered for this ailment. "Little" science—the individual doing his work by himself with just the minimum in money, equipment, and personnel that he can comfortably manage, and with a maximum amount of time to think—has not fared well. There is a legitimate place for both big and little science, but money talks, especially to directors of institutions, and the pressure on individuals to leave their ivory towers and "help support the ship" is strong.

In the last 25 years, the older brothers in academic oceanography, Scripps and Woods Hole, have been joined by a whole set of new institutions of somewhat smaller size, such as those at Columbia University, the University of Washington, and four or five others. There is also a whole host of coastal biological stations, most of which employ only a few scientists and tend to be financially insecure. The main distinction between the two categories is that the oceanographic institutions address themselves to the entire spectrum of ocean science and regard all of the world's oceans as their territory, whereas the coastal stations tend to be confined mainly to biological studies and are local in orientation. Stanford University has a marine station of this kind at Pacific Grove; like the others, it concentrates on marine animals as objects of biological research rather than on the ocean of which marine life is an integral part.

Outside the United States, active oceanographic programs are limited to a few countries because of the size and cost of the required effort, and also because a broad academic base is necessary. The United States and the Soviet Union dominate the field in terms of size, but increasingly more significant contributions are coming from other countries such as Germany, Japan, France, Canada, and the United Kingdom. None of these countries is a newcomer to the field, but in the 1950s and '60s they were dwarfed by the American effort. Now this picture is rapidly changing, and the best ideas and the most sophisticated equipment are quite frequently found outside the United States.

There are, of course, many more nations that have coasts and therefore a potential interest in the ocean. As the awareness of the resources of the sea has increased, these nations have become very politically active, but their scientific expertise has lagged far behind their needs. Because of their large numbers and hence their political power, they will greatly influence the future use of all marine resources, as we shall see in Chapter Six. If they are to play this role effectively and with solid understanding of the issues, the more advanced nations will have to furnish a great deal of assistance.

I have described the coming of age of oceanography mainly in organizational and even political terms. What about the science? Do we understand a great deal more about the ocean than we did after the *Challenger* came home in 1876? The answer of course is yes, even though we still consult the *Challenger Reports* for the remarkable depth of their insight and their precise observation. Much of our more recent knowledge details the geography of the sea; so vast an unknown territory requires much work simply to describe what is there. Even today, our descriptive knowledge is quite uneven; we know much more about the North Atlantic than about any other ocean, and very little about the southern oceans that surround Antarctica.

More exciting than the patient building of this foundation of fact have been the conceptual advances, which have occurred in all branches of ocean science. Around the turn of the century the physical oceanographers began to understand the gross features of the surface circulation of the ocean, then proceeded to develop insights into the dynamics of the flow of water at great depth. Just as they became convinced that the entire ocean circulation was understood, they discovered a whole new class of smaller (but still large) phenomena—eddies and vortices superimposed on the ocean currents much like the weather patterns of storms and high pressure cells are superimposed on the global air flow. This ocean "weather," unknown as recently as ten years ago, promises vast new insights and important practical applications, because the eddies, hundreds of miles across, affect the speed of ships, the Navy's use of underwater sound, and many other uses of the sea. The biologists, having taken an inventory of ocean life, have turned increasingly to attempts to understand the ecology of the ocean. With this has come a far better understanding of living marine resources, the reasons for the rise and fall of fish catches, and the measures needed to safeguard these resources for permanent use. The marine geologists and geophysicists, in particular, have experienced the pleasure of being at the center of a major revolution in our understanding of the earth. This revolution, fueled mainly by data from the sea, is having far-reaching fundamental and practical consequences. In the following chapters we will explore in detail many of these exciting advances in physical, biological, and geological oceanography.

It seems then that oceanographers, notwithstanding their complaints about the pressures and complexities of their current situation, have been quite successful in their quest after the mysteries of the sea and that their work will continue to have a major impact on man's well-being. Not everybody can be expected to agree with this positive view of the value of our expensive enterprises, of course, and I cannot resist quoting the following passage:

> The first volume recording the adventures of the *Challenger* yachting trip is now out and the other fifty-nine will be ready in less than a century. Everybody knows that Mr. Lowe sent a man-of-war away laden with Professors, and that those learned individuals amused themselves for four years. . . . Then the tons of rubbish were brought home and the genius who bossed the excursion proceeded to employ a swarm of foreigners to write monographs on the specimens. There were plenty of good scientific men in England but the true philosopher is nothing if not cosmopolitan; so the tax payers' money was employed in feeding a mob of Germans and other aliens. The whole business has cost two hundred thousand pounds; and in return for this sum we have got one lumbering volume of statistics, and a complete set of squabbles which are going on wherever two to three philosophers are gathered together.

This passage is not from a current newspaper, though it could well have been, but a comment on the 1872 *Challenger* expedition (cited by Margaret Deacon in *Scientists and the Sea 1650–1900*). Let us examine the squabbles of the philosophers and begin with the complete reversal in our understanding of the history of the earth that has been produced in the last 20 years, mainly through the efforts of ocean scientists, and is generally known as the geological revolution.

Memory by Elihu Vedder.

CHAPTER ONE

NO VESTIGE OF A BEGINNING, NO PROSPECT OF AN END

The prince stepped into the sea, turned around and, whispering the secret word, began to push against the shore. The island gently drifted to the other side and rejoined him with his beloved.

Old Russian fairy tale

In the second half of the 18th century, James Hutton, Scottish physician, farmer, and fine observer of the earth, came to a conclusion that sharply contradicted all then-prevailing theories and made him a founder of modern geology. Comparing rocks from ancient times with those from modern streams, volcanoes, and earthquakes, he concluded that the history of the earth can be fully explained by the processes we still observe around us every day. This insight that the present forms the key to the past differed drastically from the opinion generally held in his day that all rocks at the surface of the earth had been formed by precipitation in a primordial ocean. Known as the principle of uniformity, Hutton's concept has guided geologists ever since the early 19th century.

As everyone can see, present geological processes are very slow. Mountains cannot be observed to rise, nor is the sea normally seen to deposit its sediments on land. Consequently, if we accept Hutton's principle, we must also assume that the history of the earth has been very long. Ever since the Bishop of Ussher in 1654 determined from the Scriptures that the Creation had taken place on the 26th of October in the year 4004 B.C., it had been generally believed that the earth was about 6,000 years old. Hutton's principle was clearly incompatible with that belief, and since the beginning of the 19th century our estimate of the age of the earth has increased from a few million to the presently accepted four and a half or five billion years. However, old ideas, especially those with a religious history, die hard. From the concept of a Biblical earth complete with devastating flood came a compromise theory known as *catastrophism.* The evidence for uplift and subsidence—the marine beds exposed on mountain tops, the tilted and folded strata, and the large and often seemingly abrupt changes from one fossil assemblage into the next—all seemed to indicate a series of sudden upheavals, great floods analogous to the one named in the Bible that interrupted the otherwise tranquil history of the earth, wiped out all life, and rearranged the face of the earth.

Catastrophism, of course, flagrantly conflicted with the principle of uniformity, but it had the advantage of not requiring such a long earth history and of merely reinterpreting rather than discarding the teachings of the Church. Gradually, however, it was eroded away by increasing evidence for an ancient earth with a continuous history, and in the middle of the 19th century catastrophism received its death blow from Darwin's theory of evolution. It disappeared from serious consideration and was replaced by the modern view of an essentially endless time and infinitely slow, but very persistent, geological processes. This view, in Hutton's own words, is expressed in the title of this chapter.

Today, another fundamental change in our geological concepts is in progress. This change, modestly begun more than 50 years ago, has blossomed in the last two decades into a full-scale geological revolution and has produced the concept of continental drift and the theory of plate tectonics. The new ideas are profoundly changing the earth sciences over their entire range from petroleum geology to oceanography. They have already greatly enhanced our understanding of earthquakes and volcanoes, have exerted much influence on the exploration for oil and gas, and are about to do so in the search for ores. The main impact of the geological revolution on our understanding of earth history and the evolution of life is still to come. Even there, however, we can already trace the outline of the coming changes in the most recent views on the origin and history of the oceans and the life in them.

It is appropriate in a book about the oceans to ask why they exist at all. After all, most planets do not have seas, and the fact that the earth does is responsible for many of its characteristics, for the way life on earth has evolved, and probably for the origin of life. The question really has two parts: Where did the water come from, and what caused it to be concentrated in ocean basins rather than dispersed evenly over the surface of the earth? The first of these questions is best left to later, but the second relates closely to the new geological theories. It can be approached most easily by considering some curious facts about the nature and distribution of the oceans and continents.

Some Puzzles and Paradoxes

If we examine the distribution of elevations on land and that of depths in the sea, an interesting observation emerges. Of the approximately 30 percent of the earth's surface occupied by land, more than three-fourths lies between sea level and an altitude of two kilometers (6,000 feet). Some three-fourths of the ocean floor, on the other hand, has a depth between three and six kilometers (10,000 to 20,000 feet). Thus, the earth is essentially a two-level surface with only small transition zones and few extreme heights and deeps. This arrangement, which implies a fundamental difference between continents and oceans, was recognized more than 50 years ago.

The difference in elevation between continents and oceans tallies with a significant difference in the density of their rocks: The continents are made of relatively light material; the rocks of the ocean floor are much heavier. The two-level surface of the earth can therefore be explained nicely by assuming that both oceans and continents float freely on an underlying fluid, the lighter (and thicker) continents rising higher than the ocean floors in proportion to the difference in density between them, like so many slabs of wood and blocks of Styrofoam floating in a bucket of water. The "fluid" necessary in this model must have its upper surface about six kilometers (20,000 feet) below the ocean floors and some 30 to 40 kilometers below the mean elevation of the continents. It happens that at these depths there exists a major boundary in the earth marked by a large and abrupt change in the velocity of earthquake waves. This boundary separates the crust of the earth from the underlying mantle. Named the Mohorovicic discontinuity, Moho for short, after its discoverer, a Yugoslavian geophysicist, it was the target that the ill-fated Mohole project intended to sample by deep drilling in the ocean floor. The recognition of the Moho early in this century was an important advance that explained many observations, such as the difference in elevation between ocean floors and continents, but as we shall see it also retarded our perception of equally important boundaries situated much deeper in the interior of the earth.

Another intriguing phenomenon, known since it attracted the attention of Leonardo da Vinci, is the similarity in outline between the margins of continents on opposite sides of some ocean basins. It is most obvious in the South Atlantic, but the opposing shores of the North Atlantic can also be matched, and with some ingenuity it is possible to fit Australia, Antarctica, and India into a single block along with Africa and South America. The similarity does not stop at the geometry; geological structures, such as old mountain ranges or sedimentary rocks formed under special conditions, can be found in matching positions on both sides. In 1912, a German meteorologist, Alfred Wegener, crystallized these observations into a coherent theory of continental breakup and drift. He assumed that about 200 to 250 million years ago all continents were assembled in a supercontinent he named Pangaea, which was surrounded by a world ocean, Panthalassa. Subsequently the supercontinent broke into the fragments we know, the Indian and Atlantic oceans formed, and the drastically shrunken world ocean became the Pacific of today. Although the idea accounted neatly for a large number of otherwise incomprehensible observations, it did not become popular because nobody could conceive of a driving force that would explain how the continents could possibly make their way, like ships, through the ocean floor about which so little was known.

A third puzzling fact came to light in the 1940s and '50s. Over the years, numerous estimates had been made of the age of the oceans. The nature of the ancient sedimentary rocks on the continents clearly indicates that seas must have existed from very early in the history of the earth, certainly for the last three billion years. And yet, by whatever means geologists estimated the age of the oceans—from the salt content, the amount of sediment accumulated on the seafloor, or the oldest rocks observed in oceanic regions—the result always came to 200 million or, at the most, 300 million years, about the age of Wegener's supercontinent.

The answers to all of these and other riddles posed by the intense ocean floor studies of the last 30 years have emerged rather suddenly, primarily as a result of the evidence amassed by marine geologists and geophysicists since about 1950.

Crucial New Observations

Molten lava contains iron oxides. When the lava cools, the oxides form mineral particles which orient themselves parallel to the earth's magnetic field. Thus the direction of the magnetic field at the time the rock was formed is "frozen in," and the rock becomes a fossil compass. There is good reason to believe that the magnetic field of the earth has had a North and a South Magnetic Pole for billions of years, although the magnetic poles may not always have coincided as precisely as they do today with the poles of rotation. If we suspend a bar magnet in the magnetic field of the earth, it will not only point to the Magnetic North Pole (the declination of the magnet) but will also point downward progressively more steeply as we proceed from the equator to either magnetic pole. From this downward angle (the inclination) we can therefore determine the distance from the magnetic pole. If the magnetic pole and the pole of rotation are reasonably close together, this distance is equivalent to the latitude. The same holds true for our fossil magnets, which with the proper precautions and corrections can give us the direction to the magnetic pole and the latitude at the time the molten lava consolidated into a volcanic rock.

The study of these fossil magnets has revealed some interesting facts. If we measure the magnetism of rocks in North America that are about 200 million years old, they all point to the same magnetic pole but the position of this pole is different from the magnetic pole of today. Doing the same thing in Europe, we discover yet another position for the magnetic pole. In addition, the ancient latitude that we can derive from each magnetic rock does not correspond to its present latitude. It is not likely that the earth, 200 million years ago, possessed two Magnetic North and two Magnetic South Poles. However, if we move North America and Europe to the fit that Wegener proposed and turn them around a little bit on the globe, behold, the magnetic poles coincide with each other and the latitudes of the rocks line up in a systematic manner. If we continue our investigation on the remaining continents for the period 200 million years ago, we discover that South America, Africa, Antarctica, Australia, and India all fit together in the manner suggested by Wegener, with a single magnetic pole and a coherent set of latitudes. Next we place the two continent clusters (having fitted Asia to the Europe–North America block) on the present earth in the proper position as indicated by their ancient latitudes, and we obtain a world map of 200 million years ago (see Figure 1, next page). The quality of our information is not good enough, however, to say whether the two clusters were joined together or separated by a narrow ocean of a few hundred miles wide.

Naturally, when these results became available during the 1950s, they greatly encouraged the fanciers of continental drift, but the problem of moving the continents through the ocean basins to their present positions still seemed rather formidable. At this same time, however, the painstaking work of sea-going geologists began to reveal some intriguing secrets. First, a large mountain range was found to extend through the oceans—often rather precisely through the middle—clear around the world from the Antarctic through the Atlantic, around Africa into the Indian Ocean, between Antarctica and Australia, and finally northward

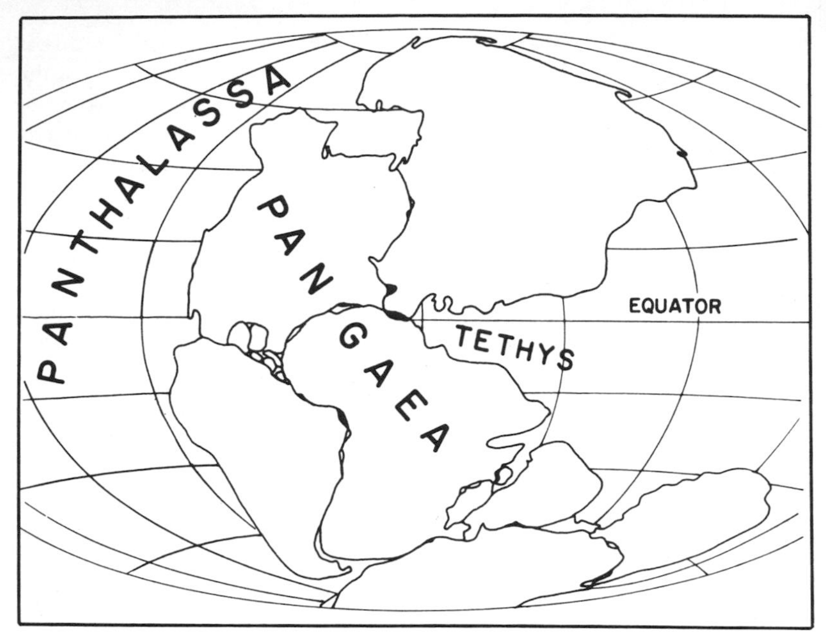

Figure 1: The world 200 million years ago as fitted together by Wegener and confirmed by paleomagnetic studies. It is possible that the great bight Tethys may have extended all the way across the waist of Pangaea.

into the Pacific where the range joins the North American continent at the Gulf of California. The configuration is similar to the seam on a tennis ball. The crest of this *mid-ocean ridge* is fairly shallow (about 9,000 feet deep) and is commonly marked by a deep cleft or rift with undersea volcanism and conspicuously fresh and young lava flows. Numerous shallow earthquakes occur along this rift, which indicate considerable tension in the crust.

Also in the 1950s we learned to tow behind our ships a magnetometer which measured deviations from the normal magnetic field of the earth. With this instrument we found that the ocean floor possessed a magnetization that was alternately higher and lower than the earth's magnetic field. These *positive and negative magnetic anomalies* formed bands parallel to the crest of the nearest mid-ocean ridge and, if colored black for a negative and white for a positive anomaly, made a magnetic map of the ocean floor look like a zebra skin. The explanation for this curious pattern was not found until six or seven years after it was discovered, and several rather exotic ideas were advanced; but finally in

1963 two young English geophysicists, Drummond Matthews and Fred Vine, put it all together.

They reasoned that if the lava of the ocean floor was magnetized during cooling in the direction of the present magnetic field, its magnetism would be added to that of the earth's magnetic field, and our instrument would see too high a magnetic value (a positive anomaly). On the other hand, if the lava had cooled in a magnetic field where the North and South Magnetic Poles were reversed as compared to the present, the opposite magnetization of the rock would be subtracted from the earth's present magnetic field and we would obtain too low a value (a negative anomaly). In the 1950s it had been discovered by means of measurements of lavas on land that the magnetic field of the earth does in fact reverse rather often, about once every 600,000 years, and the times at which some of these reversals occurred had been determined by Stanford Professor Alan Cox and his co-workers. The rotation of the earth is of course not affected by such reversals of the magnetic field; the reversals merely involve an exchange in place of the magnetic poles. There is no really good explanation *why* this happens, but that it does is well documented.

Vine and Matthews next reasoned that a series of positive and negative anomaly stripes parallel to a mid-ocean ridge crest would be just what one would expect if the volcanic rocks of the seafloor were continuously being formed in the rift on the crest, cooling, becoming magnetized in the direction of the prevailing magnetic field, and traveling away from the ridge perpendicular to the crest. In time, the magnetic field might reverse, the rocks would become magnetized in the opposite direction, move out, and so forth. Thus they saw the ocean floor as a gigantic tape recorder, recording the history of the earth's magnetic field on a tape newly formed in the rift and moving out with the seafloor. A time scale for these magnetic reversals was obtained from measurements on land and was later extended and calibrated by drilling into the ocean floor and determining the age of the oceanic rocks.

Another important piece of evidence that contributed to the geological revolution came from earthquakes. Earthquakes are far from uniformly distributed over the surface of the earth; in fact, they are highly concentrated in a few distinct, narrow zones. One of these zones is located at the crest of the mid-ocean ridge, where the earthquakes are shallow and are caused by tension in the crust. Another belt, consisting of strong earthquakes caused by compression of the earth, follows along the margins of the Pacific at the boundary between land and sea, marks many of the deep trenches in the western Pacific, continues through Indonesia and the Himalayas into the Middle East and Turkey, and traverses the entire Mediterranean. Many of these quakes are very deep, from 100 to

700 kilometers (about 60 to 500 miles) into the depths of the earth, in marked contrast with those of the mid-ocean ridge, which generally occur at depths from 5 to 20 miles. The rest of the world is fairly quiet.

The Concept of Plate Tectonics

These and other observations produced, in 1965 to 1968, what can best be described as a flash of insight that came simultaneously to several people. The concept that emerged from this is as follows. Think of the outer layer of the earth as consisting of a set of slabs or plates, perhaps 10 to 14, that are from 50 to 100 kilometers thick (50 to 70 miles) and float on an underlying material that behaves somewhat like a fluid. Note that the base of the plates, the boundary with the underlying "fluid," is much deeper than the Moho (the boundary that separates the earth's crust from the underlying mantle) which we used to explain the difference in height between continents and oceans. In fact, our fixation on the shallow Moho delayed our thinking a great deal, because if the plates are thin, the plate tectonics model will not work.

The plates are very large and they include both continents and oceans (see Figure 2). The North American plate, for example, includes half the North Atlantic Ocean as well as the continent. Plates consisting only of ocean floor are restricted to the Pacific. Now imagine a frozen river. When spring comes the ice breaks up and ice floes start traveling downstream;

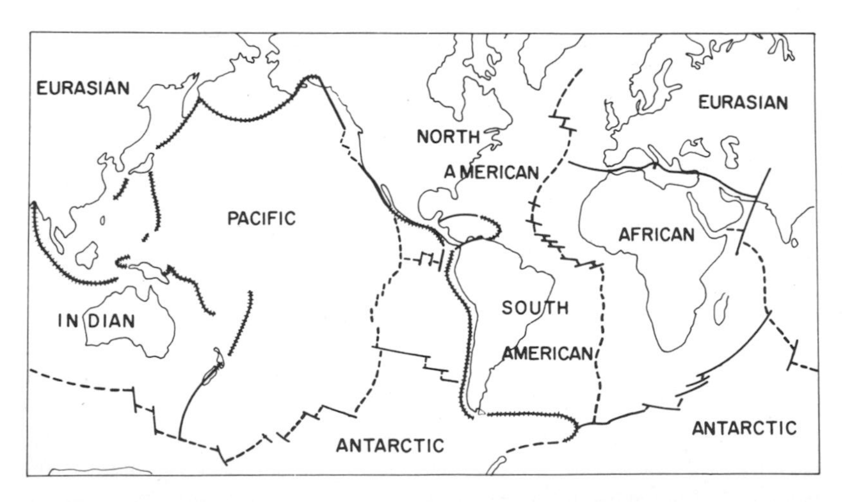

Figure 2: The tectonic plates of the world. Divergent plate boundaries are shown with dashed lines and coincide with the mid-ocean ridge. Boundaries of collision or subduction zones are marked with a hachured line: in most cases they coincide with a deep oceanic trench. The great faults or fracture zones are indicated with a solid line.

some are large and will follow the current rather closely, but many are small and are pushed both by the current and by collisions with other floes. Although the average direction of the ice reflects that of the current, it will not be easy to discern the flow pattern from the motion of individual floes. Thus we think of some sort of flow deep in the earth underneath the plates, perhaps driven by convection currents resulting from differences in temperature.

The model we are thinking of is analogous to a saucepan full of soup on a stove. In the center, the hot soup rises because it is less dense. At the surface, the rising current diverges to the sides and cools against the walls of the pan, becomes denser, and sinks to the bottom. The plates are driven by such a flow, but are affected by interactions with each other to such an extent that their boundaries and directions of motion do not allow us to discern clearly the underlying flow pattern.

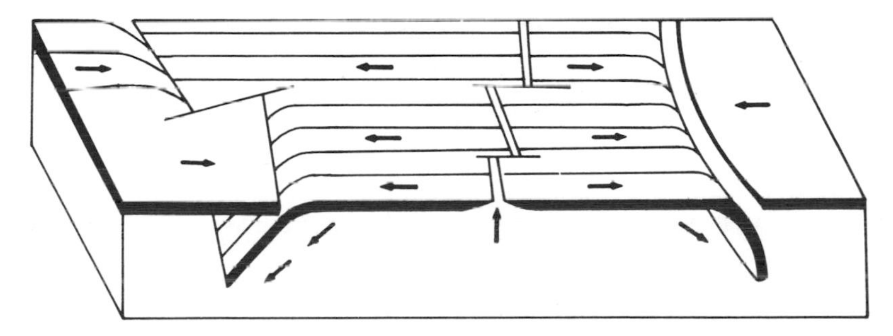

Figure 3: A cartoon of the principle of plate tectonics. The plates are indicated with the heavy black line and rest on an underlying material that behaves like a fluid. Arrows show the direction of motion away from a divergent plate boundary in the center toward collision boundaries on each side. Fracture zones offset both divergent and collision boundaries.

Three fundamental interactions between plates are possible, producing three different types of plate boundaries. Plates may drift away from each other, they may collide, or they may move alongside each other in parallel motion (see Figure 3). The different boundaries can be clearly seen on the surface of the earth. Diverging plates produce a gap where hot material wells up from the interior, cools and forms new crust which attaches to both plates, then splits and again allows new material to well up. These are the mid-ocean ridges with their rifts and the volcanism that makes new crust. Because the plates try to separate here, there is tension in the crust that produces shallow tension earthquakes. The continuous formation of new crust and its removal from the rift creates the tape recorder of magnetic anomalies, and the farther one goes away from the ridge crest, the older the ocean floor will be.

Where plates collide, there will be an excess of material that must be accommodated. If this excess were not removed, there would be no room at the mid-ocean ridge to form new crust and the motion would stop. The excess crust can be destroyed if one plate dives under the other and in that way carries the excess material back into the depths of the earth from which it came. There it can heat up, melt, and be absorbed into the earth's interior. Thus we have a full cycle from rising material at the mid-ocean ridge, lateral transport to a zone of collision, and finally return to the depths. It appears that such a cycle takes from 100 to 200 million years. The process of destruction at a boundary of collision, called *subduction,* leaves its marks in the form of a deep trench in the seafloor above the plate that goes under, whereas deep earthquakes, volcanoes, and mountains built by compression mark the edge of the overriding plate. If two plates consisting of ocean floor collide, either one of them may go down because they have equal density, but if an oceanic edge meets a continent, the lighter continent must remain on top, although it is greatly deformed by volcanism and compression and develops large mountain ranges. The South American Andes are an example of a continent-ocean collision; the curved strings of volcanic islands in the western Pacific that border deep trenches represent ocean-ocean collisions.

A second form of collision edge occurs when two continents meet. Since neither can go under, they compress each other's edges into spectacular mountain ranges, such as the Himalayas where India and Asia collide. There is a limit to the amount of crumpling that is possible, however, and the collision must eventually end. Because the two plates can no longer move, they will affect other plates on their other borders; thus the result of a continent-continent collision is often a major rearrangement of the motions of many or even all plates. Divergent boundaries may be converted into colliding ones, and directions of motion may change. Another result of collisions involving continents is that, since they cannot sink, continents are permanent, whereas the ocean floor is continuously destroyed. This solves the puzzle of the young age of the ocean floor.

Nothing so impressive happens when two plates move along each other. This kind of boundary, called a *fracture zone,* is marked neither by great heights nor great depths because all it has to accommodate are some inevitable minor irregularities. Nevertheless there is friction, which is released in earthquakes that accompany horizontal motion. Occasionally the plates may become stuck for a while, then break loose with a tremendous jar. The San Andreas fault in California and the great faults of southwestern Alaska are examples of this type of boundary between the North American and the Pacific plates, which move in opposite directions at a rate of several inches per year. The consequences are well known.

Another example is the fracture zone separating the North American and Caribbean plates, which produced the large Guatemala earthquake of 1976.

Fracture zones have an important story to tell. Because they represent a boundary of parallel movement, they tell us the direction of movement of the adjoining plates. Given enough fracture zones, and there are plenty, we can derive the directions of motion of all plates. The speed of movement we can get from our magnetic zebra stripes: If we know the width of each of the magnetic anomalies and the times when the magnetic field flipped at the beginning and the end, we can calculate the rate at which new crust was being formed and hence the speed with which the plates separated. These velocities are not small; the North Atlantic widens at a rate of about two inches per year and was about 100 feet narrower in Columbus's day than it is now. The whole Atlantic was formed in about 165 million years. In the Pacific, movement is even faster, up to 12 inches per year on the Pacific mid-ocean ridge. Crust is lost in trenches at rates that vary from 12 to 18 inches per year depending on the location.

If we estimate the rates and directions of movement for all plate pairs of the world, we can unravel the entire pattern of motion from the record of mid-ocean ridges, fracture zones, and magnetic anomaly stripes in the ocean floor. The continents, on the other hand, remain essentially unchanged except for the crumpling and modest growth of their edges. Working our way backward in time we can, in this manner, reconstruct the changing geography of the world. The technique works well to about 75 to 100 million years ago, but because the old ocean floor is continuously being destroyed by subduction, there is much more young than old ocean floor present, and eventually, going back in time, we run out of ocean floor. Fortunately at that point we are already much more than halfway back to the time when all continents were joined together. To fill the gap, we start with a good fit of the supercontinent, and using paleomagnetic measurements of latitude and orientation with respect to the magnetic pole and assorted geological information, we work forward in time by moving the continental pieces gradually apart in the appropriate directions. Somewhere along the way we meet our backward reconstruction and, presto, we have the geographic history of the world for the past 200 million years.

Consequences of Plate Tectonics

The application of these relatively simple concepts clearly shows how much the distribution of land and sea has varied in the past and how important it is to get the shifting geography of the earth straight before we can proceed to a real understanding of almost any problem of its his-

The Envoys of Moctezuma Observing the Ships of Grijalba by Diego Durán.

tory. If we desire to study the evolution of our climate from an ice age that occurred 250 million years ago through a long period of warm and globally rather uniform weather to the ice age of the very recent past (it may in fact still be with us), we need to know the positions of the continents as well as the geography and the water circulation of the oceans. When we try to understand the occurrence of oil and gas in deposits of ancient seas, we must discover where those seas were, what life existed in them that produced the raw organic matter, how they were finally filled with sediment, and what caused the strata to deform so that traps were produced.

We are also beginning to recognize that the volcanism and the upwelling molten lava from the depths of the earth at mid-ocean ridges bring with them valuable metals, which are deposited as ore bodies in the rocks and sediments of the mid-ocean ridge crests. These ore bodies then travel with the moving plate until they descend again into the interior of the earth in a subduction zone. There they are distilled and carried upward by volcanic activity into the overriding plate where we are able to exploit them in mines. Finally, the origin of life itself and its subsequent evolution are molded by the physical and chemical environment in which life occurs. These environments in turn are to a large extent shaped by the processes that create and destroy the crust of the earth and move land and sea about. The geological revolution has given us a firm basis to make reconstructions of the ancient world, and in following chapters we shall explore their implications.

All good things in life have their price, even exciting scientific discoveries. What we pay for the enhancement of our insight into the history of land and sea, climate, and resources is the addition of spatial instability to the known eternal change with time. This new degree of freedom has made it considerably more difficult to grasp the details of the world of the past, because motions on a sphere are not easily understood by human beings accustomed to visualizing the earth on a flat piece of paper. To get rid of the distortions (which can be serious in detail) quantitative methods are necessary that demand considerable effort, even though the basic concepts are simple. Geophysicists experience little discomfort from this state of affairs, but a large number of geologists are still facing their changed world with a good deal of insecurity.

Before we go on to examine some of the consequences of the new doctrine, we should turn for a moment to the world as it was before and during the time of the supercontinent and the world ocean. There are two basic interpretations of this world which, at present, cannot be resolved for lack of data. One group believes that Pangaea was the primordial continent and that the breakup some 250 million years ago was the beginning of the phenomenon we know as plate tectonics. The other view holds that the drifting of continents is a far older phenomenon and that Pangaea was preceded by a world of dispersed continental blocks and many oceans. There is some reasonable evidence that an earlier ocean preceded the North Atlantic 300 million years ago, and extreme reconstructions have divided the world of that time into numerous rather small continents or large islands which ultimately all became welded together to form the European and Asian continents. It is difficult to get a firm hold on the problem because the old ocean floors no longer exist except as detached fragments crushed against old continental margins by collisions. Thus all we have left of the old oceans is a series of deformed zones across the various continents, which may or may not represent sutures where blocks were welded together. Such presumed sutures resulting from collisions are the Ouachitas and Appalachians of North America, the Urals of eastern Europe, and the Himalayas of Asia. There is as yet simply not much agreement, even among believers in continental drift, regarding the geography of the pre-Pangaean world.

We have not yet reached the point where we understand what drives the plates—a huge convection current flowing deep in the earth, the weight of a downgoing plate in a subduction zone, or the push of material at a divergent boundary. It is obvious, however, that large plates can be driven across the surface of the earth only if they are rigid and strong, and they must therefore be quite thick. It is likely that during the first few billion years of its existence, the earth may have lacked such a thick and solid outer crust. Instead, as the lighter fractions of the planetary matter gradually distilled and floated to the surface, much like the foam on a boiling pot of soup, the first crust may have been thin, brittle, and easily fragmented—moving up, down, and sideways rapidly and more or less at random. The evidence is fairly strong, however, that by one and a half to two billion years ago, a crust had formed which, in extent, thickness, and composition, was much like the present one, so that the first of several continental drift cycles may have started then. It is also clear that during the last 200 million years the shifting pattern of continents and oceans and the volcanism and deformation that accompany plate tectonics have had a large influence on the formation of mineral resources, on climate, and on the evolution of life. It is therefore important that we eventually determine precisely the time when continental drift began, so that we understand whether similar forces operated in the millions of years before the days of the supercontinent Pangaea.

The Geological Revolution

The sudden emergence and rapid evolution of totally new scientific concepts are always enormously exciting. I remember quite vividly, having studied chemistry in high school and college as an unrelated set of compounds and reactions, the abrupt enlightenment in graduate school when understanding achieved in the early 1940s about the nature of the chemical bond placed the whole subject in a single scheme of great predictive power. The atom model of Niels Bohr in the early part of this century and the breakthrough in the understanding of the double helix in genetics in the 1950s had a similar revitalizing effect on physics and biology. Old facts acquire new meaning, jaded scientific palates are tempted, and an infinite horizon of new opportunities for investigation opens up. As a result of the plate tectonics concept it is no longer necessary to regard the geological knowledge of individual parts of the world as largely independent; instead we can devote our efforts to looking for the underlying patterns. I do not wish to imply that all the answers are in or even that the current theories are necessarily correct. There is some very valid opposition as well as some stubborn and irrational rearguard fighting, and in 20 years things may look different to us. In the meantime, however, the new ideas are immensely fertile in suggesting new problems and new avenues of research that should be examined, and we can be grateful that we live in such a time.

The rapid overhaul of ideas has created an unusual scene in the dissemination of knowledge and in the speed with which people acquire major scientific reputations. Suppose that you start a totally new branch of science and write two papers about it in the first year. In the second year your graduate student, after reading only those two papers, will be fully conversant with the field and, by writing two papers on the subject himself, will emerge as the second major expert. Even if the field catches on and the effort that goes into it doubles every year, in the third and fourth years newcomers will have to read only a few papers to keep abreast of developments and are sure to recognize the graduate student, now in the possession of a PhD, as a major authority in the field. At the same time much of the information will be transmitted by word of mouth or in other informal ways because the communication in such a small group is easy. There will therefore be no syntheses or popularizations because the practitioners have no need for them (nor the time to write them). Outsiders will soon have great trouble catching up since they are not in the communications network, and the general public essentially has no access. Rapid progress, quick fame, and an efficient expenditure of efforts thus characterize a new scientific field. On the other hand, in an old and well-established field with numerous experts only extreme specialization can keep the amount that must be read under control, and the average scientist will probably be well into his forties before a significant number of his peers will have spotted his name.

Having watched with vivid interest the only scientific revolution I have ever participated in, it has struck me, as it has others, how different the development was from the orderly progress that science is supposed to demonstrate. Between revolutions a field advances rather systematically, using observations that lead to hypotheses, which can be tested and discredited or developed into new hypotheses that lead to new pertinent observations. During our revolution nothing of the kind happened, and every rule in the book of scientific methodology was frequently and sometimes flagrantly violated. The concepts and hypotheses were often considerably ahead of a solid evaluation of the facts, and when used at all the facts were sorted in a highly selective fashion, discarding as irrelevant what did not seem to fit. There was distortion of known reality, gross oversimplification, and a great deal of emotion. Once, after having expressed some strong skepticism regarding one of the far-out ideas, I walked into an elevator at a meeting of the American Geophysical Union. Two protagonists of the idea followed, stopped at seeing me, and walked off again, obviously not intending to share the close proximity of such a conservative sorehead. Trust in friends, distrust, words like "he is a first-rate man" or a "sloppy investigator" assumed a great deal of importance in judging the pros and cons.

Revolution seems certainly the appropriate word for the goings-on, even though the scientific concepts were not totally new. A period of consolidation must follow. We are probably in it now. Soon the new paradigm will be the conventional wisdom. There is, inevitably, a sense of being let down at such a time; for some young scientists who made a major name for themselves almost overnight, there is the problem of an encore. It is more pleasant to present great ideas than to engage in painstaking collection of data. Nevertheless, in the next decade, this daily grind is likely to produce some striking and far-reaching new insights.

Nocturne in Black and Gold: Entrance to Southampton Water by James A. McNeill Whistler.

CHAPTER TWO

THE CHANGING FACE OF THE SEA DIMLY OBSERVED

> *. . . For that my opinion dooth differ from some of the auncient writers in natural Phylosophy, it is possible that it may be ytterly dislyked of and condemned to be of no trueth.*
>
> William Bourne, *The Treasure for Travellers,* 1578

THE BREAKUP of the supercontinent 200 million years ago and the subsequent drift of the fragments produced large changes in the geography of the earth. As global seaways opened and closed, permitting water to flow where it once was blocked by land or deflecting currents, the circulation of the oceans was also modified. Climate, although mainly a function of the incoming solar radiation and hence closely related to latitude, is also strongly influenced by the size and shape of the continents and by the oceanic circulation, as can be seen when comparing the weather in the interior of North America with that of the coasts, or the climate in Labrador with that in England, both at the same latitude. Thus the evolving geography of continents and oceans is certain to have been accompanied by climatic changes. Finally, because life closely adapts to its environment, the drift of the continents must be considered a potential influence on biological evolution, perhaps a major one. Our understanding of all of these aspects of drift is in its infancy, but the outline of insights soon to come can be seen in the results of studies, begun four or five years ago, of the ancient ocean. Before we can discuss the history of the ocean, however, we must first refine our understanding of the changing shape of the ocean basins and examine the shifting boundary between land and sea.

Reconstruction of Old Oceans

The mid-ocean ridges are the most prominent topographic elements of the ocean floor. There molten rock wells up from the depths of the earth, and heat is transferred from the earth's interior to the ocean floor. Liquid lava or hot rocks have a greater volume and are therefore lighter than the same substances at a low temperature. As a result a hot ocean crust floats higher than a cold one—hence the mid-ocean mountain ranges. A continent heated from below must also rise; the Colorado plateau is high because the crust underneath is unusually warm and therefore has a low density. In the Gulf of California, a mid-ocean rift on the boundary between the Pacific and North American plates penetrates into the continent. The crust in the Gulf is hot and light and has caused the continental blocks on both sides to rise and form striking mountain ranges.

As the new crust moves away from its place of origin, it gradually cools and becomes denser. It must therefore sink, with the rate of sinking proportional to the rate of cooling. If we plot the depth of the ocean floor against its age, we find that the curve at first descends steeply from the average depth of the crest of the mid-ocean ridges, which is about three kilometers (10,000 feet), but after 10 to 15 million years the subsidence slows down. Crust 75 to 100 million years old, having reached a depth of about six kilometers, sinks no further. Consequently the deepest parts of the Atlantic are also the oldest and the most remote from the Mid-Atlantic Ridge. The oldest Pacific floor, in the extreme west of the ocean, is about 150 million years old and is six kilometers deep. Sediments that have accumulated since the crust was formed may decrease the depth slightly, but except near land the effect is small.

Because of the simple relationship between the ocean floor's age and its depth, we can reconstruct the relief of the ocean floor in the past by taking the present relief and applying corrections to it. The results are always interesting and sometimes striking. The present South Atlantic has a distinct mid-ocean ridge, the Mid-Atlantic Ridge, as well as deep basins on either side between the ridge and the continental margins (see Figure 4). This rather simple relief is complicated only by two shallow plateaus that extend seaward from southern Brazil and southern Angola about two-thirds of the way down from the equator. These two plateaus, the Walvis Ridge and the Rio Grande Rise, are separated from the crest of the Mid-Atlantic

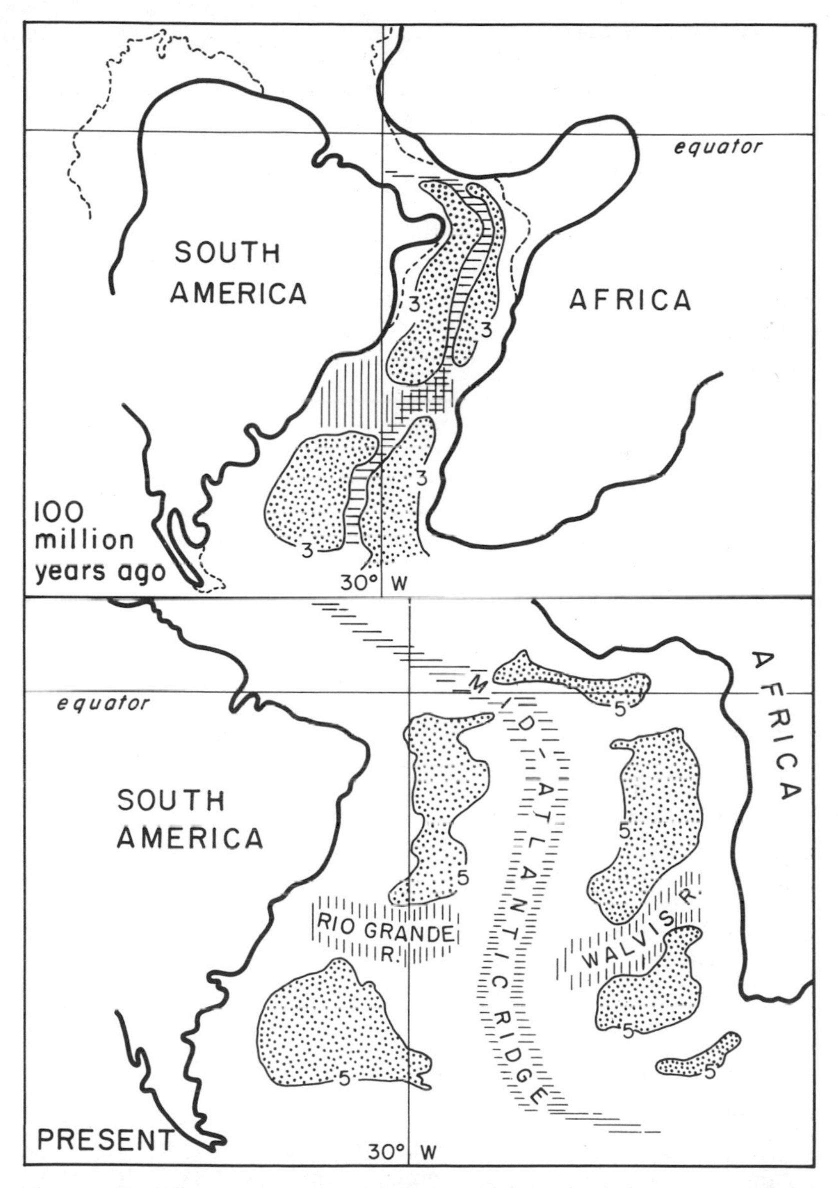

Figure 4: The changing geography of the South Atlantic. The three major topographic features are the Mid-Atlantic Ridge (shown with horizontal shading), the Rio Grande Rise and Walvis Ridge plateaus (vertical shading), and the old and deep basins (stippled). The Rio Grande and Walvis plateaus were joined together 100 million years ago to divide the South Atlantic into a northern and a southern basin, whereas today they are separated from each other and the Mid-Atlantic Ridge by deep water passages. Note the difference in shoreline of the continents between the two periods shown (the present shoreline is dashed on the 100 million year map). Note also that the deepest basins were only 3 kilometers (10,000 feet) deep 100 million years ago, but are more than 5 kilometers (17,000 feet) deep today.

Ridge by deep troughs and do not block the flow of cold, deep water from the southern ocean northward into the North Atlantic.

If we reconstruct the South Atlantic of the remote past, we discover that 100 million years ago this ocean basin was not only very much narrower, but also was divided into a southern and a northern basin by a shallow transverse barrier (consisting of two plateaus studded with islands or perhaps even mostly above sea level) which connected southern Brazil with Africa. As the two continents continued to drift apart, the barrier broke and sank, and its remnants are now the Rio Grande Rise and the Walvis Ridge, both about two kilometers (6,000 feet) deep. While the barrier existed, it not only prevented all flow of water between the northern and the southern basins except at very shallow depth, but also provided a pathway from one continent to the other along which shallow-water and island-hopping organisms could migrate. The northern basin remained essentially landlocked until about 80 million years ago, with only shallow and restricted passages in the north and south to the open sea.

Landlocked seas have properties very different from those of the open ocean. If they are located in a warm climate, they can easily become highly saline because of evaporation. On the other hand, if there is considerable runoff from the adjacent land, they may become brackish or even fresh. The extremes in salinity sometimes alternate in time, making the basin a difficult environment for life. Environments of such high stress usually produce communities consisting of few species but large numbers of individuals. The Mediterranean is a good example of a landlocked sea that has experienced extreme environmental changes. About six million years ago, the Mediterranean (then completely enclosed) evaporated totally, and thick salt beds were laid down. Shortly thereafter the Atlantic Ocean, probably quite abruptly, broke through the Straits of Gibraltar and refilled the basin with seawater. For several million years the Mediterranean environment was probably rather like it is today, but when the Ice Age arrived in the Northern Hemisphere a few million years ago, the rainfall and the runoff increased a great deal, making the Mediterranean considerably fresher than it is now. Today, because of evaporation and little runoff, it is somewhat saltier than the Atlantic.

Landlocked basins may also be supplied with an exceptionally large amount of nutrients from the adjacent land; these nutrients increase the fertility and the production of organic matter. The bottom, which is not continuously flushed by deep currents, may at times become stagnant so that the large quantity of organic matter settling from the surface is almost completely preserved, thereby forming incipient source beds for petroleum. During its early history, the South Atlantic demonstrated all of these effects. About 110 million years ago it was highly saline—among its earliest sediments are thick beds of salt. A few million years passed, the ocean widened, the Rio Grande and Walvis barriers sank somewhat, and conditions became less saline. Life flourished and produced abundant organic matter which was well preserved in the still poorly ventilated deeps. Finally, when conditions became like those of the normal ocean, fine-grained oceanic sediment covered the black oozes, rich in organic matter, which then simmered for millions of years to produce large oil reserves.

Shifting Shorelines

Up to this point we have tacitly assumed that continents are always synonymous with land and ocean basins with sea. This assumption is not strictly valid: The continental shelves and the Gulf of Maine are examples of seas on continents. Moreover the boundary between land and sea has shifted across the continents in the past, as anyone knows who has found marine fossils in sediments on land. We have long known about the various encroachments of the sea onto the land; Leonardo da Vinci speculated on their causes in a surprisingly modern fashion. Sea level changes were originally explained by assuming that the land rises and sinks over broad areas as a result of mountain-building forces. Gradually, however, as geologists became more adept at assigning precise ages to rocks, they discovered that some of these shifts of the shore were global—not merely local—and thus could not be comfortably explained by local causes.

The first known global sea level changes were associated with the locking up or releasing of large volumes of seawater in ice caps during the last ice age. The earth has known several ice ages, the last of which began in the Southern Hemisphere about 10 to 15 million years ago though its origins can be traced to 40 or 50 million years ago. An ice age consists of periods of varying severity, the last of which began two or three million years ago; each period itself includes advances *(glacials)* and withdrawals *(interglacials)* of the ice. During each glacial a large drop in sea level occurs. At the peak of the most recent glaciation, about 20,000 years ago, the shoreline was located more than 300 feet below the present sea level. In California, a drop of 300 feet (about 100 meters) below present sea level does not increase the land area much because the seafloor is steep, but in the Gulf of Mexico the 100-meter depth contour lies from 100 to 200 miles out at sea. When the ice began to melt 15,000 years ago, the sea recovered its territory very rapidly, initially at a horizontal rate of more than 100 feet per year. Beach real estate must have lost much in value in the process. If all of the present ice on Greenland and Antarctica should melt, the sea would rise by another 200 feet, completely drowning many, even most, of the large population centers of the world, including New York, Amsterdam, and Calcutta.

Many other synchronous global sea level changes are now known, but few can be explained by glaciations. In fact, over the last 600 million years, although only two or perhaps three ice ages have occurred, the sea level has risen slowly and fallen abruptly many times. In the beginning each of these cycles of slow rise and rapid drop took about 50 million years; more recently, however, they have speeded up to a peak about once every 10 to 20 million years. Currently the sea level is low, but in another 10 million years there will be a good deal less room on the continents. Superimposed on this sawtooth pattern of rise and fall is a slow, long-term rise which began 500 million years ago and reached a peak about 75 million years ago. Then slow decline set in which continues today.

The peak of the long-term rise coincided with a major short-term increase, and 75 million years ago a large part of the continents—perhaps as much as 40 percent of the present land area—was flooded. In North America, the sea extended over the entire mid-continent region from the Gulf of Mexico deep into Canada and from the Rocky Mountains to the Appalachians. Worldwide, the extent of shallow seas was quadrupled compared to the present. Africa was split into two large islands by a sea extending from the Gulf of Guinea to the Mediterranean. Although the other short-term rises were less spectacular than the one of 75 million years ago, each brought about drastic changes in the outline of the land masses, temporarily at least doubling the area covered by shallow seas. The effects of these fluctuations for climate and life were, as we shall see, profound.

It is not intuitively obvious that changes in sea level, global though they may be, have anything to do with plate tectonics, which involves after all mainly horizontal motions. Yet if we cannot call upon the ice ages for an explanation, only two possible causes remain: changes in the amount of seawater on earth or changes in the capacity of the containers, the ocean basins. Changing the volume of seawater appears impossible. Even if we could think of a good way to release suddenly a large volume of water from the interior of the earth, so that sea level would rise, where could the water go when it was time for sea level to fall again? Thus we must look for changes in the volume of the containers, the ocean basins.

Earlier in this chapter I explained that the depth of the flanks of a mid-ocean ridge depends solely on the temperature of the crust. The older the crust the cooler, and the deeper the seafloor. Accordingly the depth depends only on the time elapsed since the crust was formed. If the plates move apart slowly, little new crust will be added each year, and a short distance away from the plate edge the seafloor will already be cool and deep. Therefore the mid-ocean ridge will be steep and narrow. If, on the other hand, the plates separate rapidly, the crust quite far from the plate edge remains young, hot, and shallow, and the ridge will be broad and

In Polar Seas by William Bradford.

gentle. Because the combined volume of all mid-ocean ridges is a significant fraction of the total volume of the ocean basins, changing narrow ridges into wide ones, or vice-versa, significantly changes the total volume of the oceans and decreases or increases the amount of water that they can contain. If we speed up the rate of plate motion, sea level must rise; if we slow it down, sea level will fall. Similarly, changing the number or the aggregate length of mid-ocean ridges would also change the volume of the ocean basins. Thus if the number of plates increases, the continents will be flooded. Whereas a supercontinent in a single world ocean is likely to be mostly above water, a world of fragmented continents will be endowed with many mid-ocean ridges, intermittently flooded land masses, and many shallow continental seas.

The idea that variations in the rate of continental drift would be responsible for changes in sea level is simple and quite logical. It also provides a good explanation for the great flood of 75 million years ago, because there is reasonable evidence that the movement of the plates at that time was more rapid than either before or after. Unfortunately, we cannot demonstrate that plate movements correlate with the many other fluctuations of sea level over the last 500 million years. For most of that time there is simply not enough information on the rates of plate movement or on the number and length of mid-ocean ridges, whereas for the last 100 million years—for which the data are better—the information is quite ambiguous and does not really support the correlation between plate movements and sea level changes. On the other hand, we have no other acceptable hypothesis to explain the remarkable rise and fall of sea level over the ages and are rather like the little girl I once met on the beach in La Jolla who asked me, watching the tide go out: "Where does all the water go?"

The Currents of the Sea

The ocean is not a still body of water. Besides the tides and waves there are large current systems that circulate the water, both in a shallow surface layer a few hundred feet thick and in great depths. The surface circulation is driven primarily by the wind. Because a steady breeze from one direction will move more water farther than will a series of short-lived gales from various directions, the main planetary winds are the ones that drive and shape the circulation patterns of the oceans. These are the winds resulting from the rotation of the earth—the easterly winds of the polar regions, the westerlies of the mid-latitudes between 40° and 60° north and south, and the southeast and northeast trades between 10° and 20°—which are separated by zones of calm or of varying winds.

The trades drive the tropical waters westward with a return flow along the equator. Much of the water, however, piles up against the continents at the western boundary of each ocean and is diverted to the north and south until it reaches the latitude of the westerlies, which blow it back across the ocean and return it to its point of origin. Thus the circulation in the northern and southern half of each ocean resembles a gigantic eddy or gyre with a quiet center where water moves neither in nor out. Because moving water is essential to supply nutrients to marine life, these central gyres are veritable marine deserts. On the other hand, the shear between opposing currents along and on both sides of the equator stirs the sea to bring up nutrients from the deep that replenish those consumed by the organisms living near the surface. This process of *upwelling* makes the equatorial current belts some of the most fertile regions in the world and the locale of major fisheries.

There is a curious force, the bane of beginning oceanography students, that affects all currents on earth. It is called the *Coriolis force* and is caused by the rotation of the earth. On the Northern Hemisphere the Coriolis force deflects a current of fluid or air to the right (looking downstream) and on the Southern Hemisphere to the left. It is said that the same force causes American bathtubs to empty with a clockwise eddy while in Australia the water gurgles out in the opposite direction (but experiments have failed to confirm this). Because a north equatorial current must diverge to the north and a south equatorial current to the south, the zone between them will require some additional water that wells up from below and supplies more nutrients. The same Coriolis force explains why in the Northern Hemisphere the wind circles clockwise around a high pressure area and counterclockwise around a low pressure storm center. Thus it is the southwest wind that brings rain to California from a storm moving in from the north.

There is one more force that drives ocean surface currents—gravity. At the equator the ocean receives far more heat from the sun than at the poles. Because warm water is less dense than cold, the sea surface at the equator is slightly higher than at the poles. Gravity drives the warm water north and south down this slight slope, but the Coriolis force converts these northward and southward flows into eastward currents which enhance the currents driven by the westwinds at mid-latitudes. Without the Coriolis force the warm water would move much farther north before cycling back, which would to a considerable extent reduce the north-south temperature gradients.

If the world were totally covered with ocean, we would have three major globe-girdling current systems: a complex system at the equator consisting of two zones of westward flow and a return flow in the middle, an eastward flow at mid-latitudes, and a westward flow in the polar regions. On the real earth, continents stand in the way of such an ideal pattern except around Antarctica where the circum-Antarctic current proves that the concept is sound. Elsewhere each current eventually bumps into an opposing continent or group of shoals and islands and is deflected to the north and south. In the Pacific the deflection of the equatorial current is mainly to the north, driving the water past Japan, which is warmed by its tropical temperature. Then, gradually cooling, the current flows east across the Pacific with the westwind drift and, meeting the coast of North America, is forced southward to its point of origin in the eastern tropical Pacific. The Gulf Stream in the North Atlantic, in the same manner, derives its water from the Atlantic north equatorial current and from the balmy seas of the Caribbean, flows north along the coast of the United States and then northeast, warming northwestern Europe and providing it with a climate that is most pleasant compared to that of Labrador at the same latitude on the opposite side of the Atlantic.

There are, of course, many significant variations on the basic pattern of oceanic circulation that depend on local factors. Nevertheless, if we assume that the rotation of the earth has been constant over geologic history and if we have a reasonable geography of land and sea for any given time in the past, we should be able to reconstruct the surface circulation of the ocean with some confidence. It would help if we had some idea of the distribution of solar heat across the latitudes because that not only controls the temperature of the sea surface and influences the flow, but also determines the position, width, and intensity of the major wind systems. That sort of information, however, is not handily available, and we are forced to examine the distribution of fossils and characteristic sediment types in order to test a series of possible circulation models with various temperature distributions. If lucky, we may derive the temperature distribution itself in this way and thereby make an important contribution to the global climatic history.

A Short History of the Ocean Circulation

Let us trace, by way of an example, the changing arrangement of oceans and continents since Pangaea fell apart and derive from that how the surface circulation may have changed in response (see Figure 5). About 175 million years ago, not long after Pangaea's fragmentation and the beginning of continental drift, the dominant ocean circulation was probably still quite similar to that of the preceding world ocean. If, for convenience, we ignore the polar flows and the internal complexities of the equatorial system, we can surmise the existence of two very large gyres, one in each hemisphere. They may have been connected by a seaway, the Tethys, that cut through the heart of Pangaea approximately along the equator, but the connection was probably not very significant. Geological evidence indicates that the climate then was more uniform than today because fossil assem-

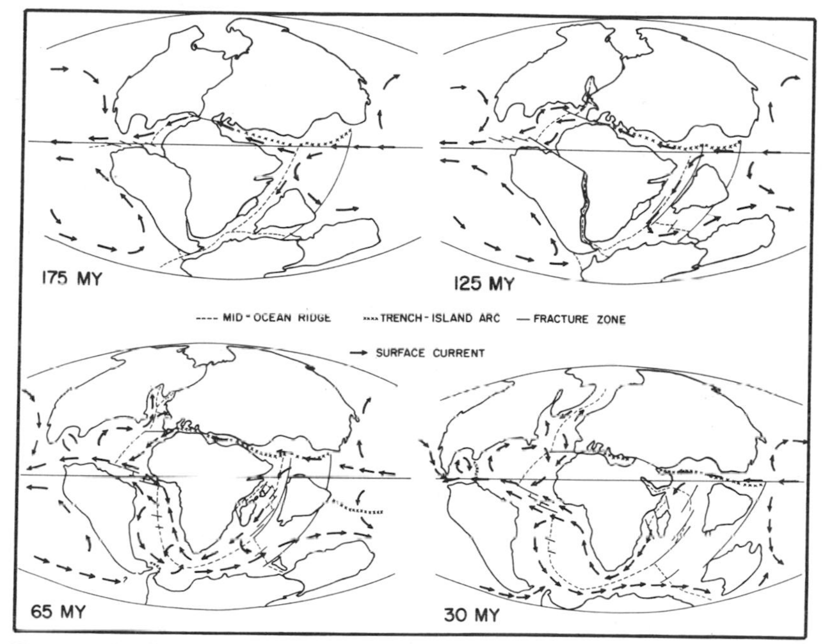

Figure 5: The changing geography of the world over the last 175 million years (MY in the figure) and the oceanic current systems that probably resulted from the interplay between wind systems and opening and closing seaways. The nature of the plate boundaries is also shown with symbols: trenches and island arcs represent plate collision boundaries. See the text for a discussion.

blages from different ancient latitudes do not show the striking differences that today distinguish the polar from the equatorial fauna. If there was indeed a smaller temperature gradient from the equator to the poles, the westwind drift would have been located at a rather high latitude, and the gyres must have been very broad and would have aided in distributing the sun's heat more evenly over the surface of the earth.

Some 50 million years later the continents were noticeably more dispersed, although the northern North Atlantic was still closed to the Arctic, and Africa, South America, Australia, and Antarctica were joined in a single block. A globe-girdling equatorial current system was possible at this time because the Tethys had become much wider. It must have been quite a current with nothing to stop it on its way, but ingenious laboratory experiments in a spinning dish full of water with model continents and with fans to provide the wind suggest that the equatorial current system was broad rather than fast. Because there were no longer any barriers across the equator, less water was probably diverted to the northern and southern gyres than had been the case 50 million years earlier. This may have caused the gyres to be narrower, pushing the westwind drift closer to the equator, thus starting a trend toward a more strongly zoned climate distribution. On the other hand, the uninterrupted flow around the equator insured that the water remained for a long time in the hot equatorial region. Consequently the rather small quantity diverted northward and southward was probably very warm. It is therefore likely that the temperature distribution from pole to pole still remained quite even. In the nascent Indian Ocean, India had moved out to the north somewhat, and a branch of the equatorial current surrounded it with warm tropical water that certainly made the southern Indian Ocean a good deal warmer than it is today.

The next stage, 60 million years ago, showed more drastic changes. All oceans were open, although the Arctic was not yet connected with the North Atlantic, and Australia was still attached to Antarctica. Equatorial water now traveled from the Indian Ocean to the Atlantic by two separate paths. A short path through the Tethys allowed warm water, with its Pacific and Indian fauna and flora, to enter directly into the central Atlantic and then travel to the Pacific. A longer path around Africa involved more cooling and mixing with Antarctic water and must have cooled the southern Atlantic considerably. The result was no doubt a marked increase in the zonality of temperature in the "Eastern" Hemisphere, if one wants to call it that (we have no good idea of longitudes). This is confirmed by the first appearance, in Europe, Africa, and the Americas, of distinctions between northern, tropical, and southern fossil assemblages.

Finally, 30 million years ago the world began to look a bit like the present one. A seaway had opened up around the world in the far south, and a cold current had begun to circle Antarctica. Because there were no barriers in its way, relatively little water was diverted northward, the current became very cold, and the temperature contrast from the equator to the South Pole gradually increased. Particularly in the South Atlantic, the appearance of a cold circumpolar current must have cooled the region in the mid-latitudes quite drastically. Moreover, the equatorial current system, rather than continuously circling the earth, had fragmented into several separate segments, each of which remained in the tropics for a shorter time than before and diverted more water to the north and south. Thus something similar to the present climate zones was developing. There is solid geologic evidence that Antarctica, which had gradually drifted south and was now approximately centered on the South Pole, had acquired glaciers and perhaps even the beginning of an ice sheet extending all the way to the coast. Because the circum-Antarctic current kept the water cold, the ice could advance out to sea and, in turn, make the current colder still. Thus the process kept reinforcing itself until 10 to 15 million years ago the first major Antarctic glaciation appeared, signaling the onset of the Ice Age that lasted at least until the last glaciation 20,000 years ago. Since it is possible that the present period of milder conditions is just an interglacial, we may well still be in an ice age. The Arctic Ocean, which had long been kept warm by water coming through a deep connection with the Pacific, did not become covered with ice until about five million years later.

Thus the slow climatic deterioration of the last 100 million years was much enhanced, although probably not entirely caused, by the gradual rearrangement of the continents and the consequent changes in the surface circulation of the ocean. All that now remains is to prove this story and to supply the details with the aid of the climatic and oceanographic records preserved in the sediments and fossils of the ocean floor. During the last eight years the deep-sea drilling expeditions of the *Glomar Challenger* have provided a considerable quantity of vital information for this job that should keep me and my colleagues happily and usefully occupied for years.

The Deep Waters of the Oceans

The circulation of oceanic surface water is not the only element of interest in a study of ancient climates. The deep water of the ocean, below 1,000 meters (3,000 feet), also circulates, although much more slowly than the surface layer. If this were not so, the ocean depths would be stagnant, the sinking organic matter would deplete the oxygen, and a smelly environment of hydrogen sulfide and other noxious gases would eliminate all life in the abyss. The gradual flow of heat from the interior of the earth would slowly warm the bottom water until it became warmer than that at the surface. Because warm water is less dense than cold, the unstable ocean would eventually turn over and vent the whole mess to the surface, with catastrophic effects on the fauna and flora of the surface waters and perhaps some poisoning of the atmosphere.

The circulation of deep and bottom water depends on density variations, with denser water sinking under the less dense. Density is controlled by salinity and temperature. A high salinity, resulting from evaporation, produces dense water and cooling does this even more effectively. Cold water is formed in polar regions, high salinity water in tropical latitudes. The formation of sea ice also increases the salinity because the salt is expelled into the underlying water as the ice forms.

At the present time the bottom water of all oceans comes from the circum-Antarctic region where low temperatures and the formation of sea ice produce dense water that sinks to the seafloor and flows northward. At intermediate depths, between 1,000 and 4,000 meters, the situation is more complicated. In the Atlantic, cool water of medium salinity is formed just south of Greenland and flows south well below the surface layer all the way to the southern ocean, whereas in the Pacific the intermediate flow is in part driven by high salinity water formed in low latitudes. In principle the deep circulation of today is therefore produced by the strong climatic zonation of the earth, and the sources of the deep and bottom water are mainly in the polar regions. From its source, the bottom water takes more than a thousand years to travel to the opposite end of the ocean and return to the surface. As it begins to sink, this deep water is rich in oxygen, but as it travels it gradually becomes depleted in this life-supporting element.

In a world without pronounced climate zones and polar glaciations, such as existed 100 million years ago, where would the bottom water come from? Must we consider the possibility of a stagnant ocean? At the present time, this question is unanswerable. There is some indication that during the period from 80 to 110 million years ago large parts of the ocean floor were poorly supplied with oxygen. Black, highly organic muds, devoid of the remains of bottom-dwelling organisms, covered much larger areas than in the present ocean and have become a rich source of petroleum. These muds are not extensive enough, however, to assume that the world ocean was wholly stagnant.

Can we think of other sources of dense water? During periods of extensive flooding of the continents, the shallow seas, if located in warm latitudes, may have been a source of high salinity water as a result of evaporation. If the salinity were high enough and the bottom water of the ancient ocean not too cold, such water could have been flushed down into the abyss. A modern example is the Mediterranean, which discharges water of moderately high salinity into the Atlantic where it sinks to mid-depth and can be found almost as far west as the Windward Islands of the Caribbean. In an ocean without Antarctic bottom water, the Mediterranean high salinity water might sink all the way to the seafloor. Similarly, extensive shallow seas in higher latitudes may cool enough in the winter to form dense cold water—as happens today in the Sea of Okhotsk in Siberia.

Thus in the days before the Antarctic was glaciated, there may have been various sources yielding different kinds of bottom water—some cool, some salty—leading to much more localized patterns of bottom-water flow than exist today. In some parts of the ancient oceans or at certain times, stagnant conditions may have occurred, and even occasional overturning of parts of the ocean when the density stratification became unstable is not inconceivable.

The composition and behavior of the bottom water in the remote past are of considerable interest. The bottom water plays a role in the formation of source beds of oil and, if sufficiently different in composition from the water of today, may have done so for marine minerals as well. The changing conditions at the seafloor may variously have improved conditions for life or threatened the existence of entire groups of animals. Finally, the bottom water plays a major role in regulating the supply of calcium carbonate, which is important in oceanic life.

Bottom Water, Calcium Carbonate, and Life

The surface waters of the oceans are supersaturated with calcium carbonate. Out of this material many marine organisms, both plants and animals, build skeletons and shells, which settle to the bottom as sediment after the organisms die. These organisms, primarily microscopic ones, use the material so freely that their annual production of shells is six times larger than the amount of calcium carbonate that the rivers supply to the oceans. The biological consumption of calcium carbonate can continue indefinitely only because there is a clever recycling mechanism which, up to a point, adjusts the supply so that it always exceeds the demand. The microscopic shells sink through the water, and if the seafloor is not too deep (above 10,000 feet or so), they arrive on the bottom essentially intact and form sediment. Below that level, however, the water is progressively more corrosive to calcium carbonate, and as the seafloor deepens, more and more of the carbonate on the bottom is dissolved shortly after arrival and put back into solution to be recycled to the users at the surface. At the greatest depths, below 15,000 feet, the supply from above and the dissolution at the seafloor are equal, and no carbonate sediment is permanently deposited.

We can visualize the ocean floor as mountain country in the winter, with snow composed of the white calcium carbonate shells in the higher elevations and bare ground below, separated by a "snow limit" which is one of the most distinctive features of the oceanic sediment cover. About four-fifths of the ocean floor lies below the snow limit. Now, if the consumers at the surface should suddenly decide to double their rate of reproduction and therefore need twice as much carbonate, the ocean would respond by dissolving more of the material on the deep-sea floor and recycling the carbonate. The effect on the undersea mountains would be immediately visible as a rise in the snow limit to higher elevations. The adjustment could continue, if necessary, until all carbonate shells were dissolved immediately after they had been produced.

We can combine the understanding of the flow of ocean water, gained from our analysis of the shifting pattern of continents and oceans, with information furnished by changes in carbonate dissolution to sketch a more detailed history of the evolution of the bottom-water circulation of the oceans.

Origin of the Deep Circulation in the Oceans

The present flow of bottom water starts in the southern circumpolar regions with the formation of cold and dense surface water, which sinks and moves northward at great depth in all oceans. Basically, the deep circulation system is thus driven by the glacial conditions which at present prevail in Antarctica, without which the present deep circulation would not be possible. When did this circulation system begin and how did it evolve? We can approach this question along several paths.

A first line of evidence is the gradual change of the continents as they drifted apart during the last 100 million years. As late as 50 million years ago, there was no free passage of water around Antarctica because South America and Australia were still attached to the southern continent. Consequently the water driven east by the wind along the southern edges of the oceans stayed at these cold latitudes only for a short time before it encountered an obstacle and had to return north to warmer climes. In contrast, water could freely flow around the world at the equator and thus stayed in the hot low latitudes for a long time. Gradually this configuration changed, first by the opening of a seaway between Australia and Antarctica 35 to 40 million years ago, then by the opening 25 million years ago of the Drake Passage between Antarctica and South America. When the Drake Passage reached its full width about 15 million years ago, the path was free for the present vigorous circumpolar current, and the long residence time of the ocean water in high southern latitudes caused it to cool considerably more than it had previously. During the same period from 10 to 40 million years ago, the equatorial seaway about the world gradually closed, and the residence time of equatorial warm water was diminished.

The result of such major changes in global ocean circulation was a large increase in the temperature gradient from the equator to the poles, which transformed an earth with relatively indistinct climate zones into the present one with distinct tropical, subtropical, temperate, and polar climates. We do not yet know whether this rearrangement of the oceanic circulation was sufficient to produce this climatic change, but without doubt it was a major contributing factor. As a result the first mountain glaciers, which had shown up in Antarctica about 50 million years ago, reached the sea 10 million years later as is shown by the sudden appearance of ice-rafted pebbles in the sediments of the southern ocean. At that time Antarctica probably looked rather like the Pacific coast of Alaska does today.

The arrival of glaciers at the seashore and the increasing current that resulted from the separation of Australia and Antarctica produced a drastic cooling of the waters of the southern ocean around 40 million years ago. Nature has provided us with an elegant thermometer for measuring ancient temperatures. The oxygen in seawater consists of two isotopes: the abundant isotope (^{16}O) and a rare, slightly heavier isotope (^{18}O). When organisms extract calcium carbonate from the sea for their shells, they pick up these two isotopes in a ratio that depends directly on the temperature (as well as some other factors, such as salinity, for which we can make corrections). Thus by measuring the $^{18}O/^{16}O$ ratio in fossil shells of organisms that lived either at the sea surface or on the ocean floor, we can determine with fair precision what the water temperature was at the time they were alive. This thermometer has told us that 50 million years ago the surface water around Antarctica was a comfortable 70 degrees Fahrenheit (20 degrees Centigrade). The temperature subsequently dropped—first slowly, then rapidly—until shortly after 40 million years ago it reached 50 degrees. Simultaneously the bottom temperature, which was a high 60 degrees 50 million years ago, fell—first slowly, then sharply—to reach 40 degrees 40 million years ago and then gradually sank to the present 36 degrees. A sharp drop in both the surface and the bottom temperatures occurred during the period 10 to 15 million years ago when the opening of the Drake Passage allowed the circum-Antarctic current to reach its present extent and velocity.

The establishment of the present Antarctic circulation had two major consequences. The circum-Antarctic current, like other major currents, vigorously stirs the surface waters of the ocean and hence is very fertile. In these cold latitudes small plants with siliceous shells, the diatoms, are particularly adapted to take advantage of the enhanced fertility, because they flourish even with large differences in sunshine between summer and winter. As a result they are abundant, and their shells dominate in the sediments of high latitude seas. The rather sudden development of the circum-Antarctic current is marked in the sediments of the southern deep-sea floor by the appearance of deposits consisting almost entirely of diatom shells.

A second consequence of the establishment of the circum-Antarctic current was much more extensive and can be observed in oceanic sediments as far north as the equatorial Pacific. For millions of years, the deep circulation in all oceans had been very sluggish, perhaps almost stagnant. Rotting organic matter and high bottom-water temperatures combined to make the seafloor environment very corrosive for calcium carbonate. The "snow limit," the boundary on the seafloor that separates the white calcareous sediments of the mountains and shallows from the brown muds of the abyss, was located at a very shallow depth. About 35 to 40 million years ago, the formation of water around Antarctica cold enough to sink set in motion a precursor of the present bottom circulation. The flow of this cooler and denser water swept out the old, stagnating, corrosive bottom waters of the equatorial region and, because it was much less corrosive, caused the snow limit to drop almost 1,500 meters. As a result, white calcareous sediments covered large areas of older brown muds.

Much later, when the circum-Antarctic current became fully established 10 to 15 million years ago, the snow limit rose once more. From this one must conclude that these intensely cold waters were heavily charged with carbon dioxide and hence very corrosive. The rise of the snow limit at that time marked the real beginning of the Ice Age, and that first phase was one of the most severe. The story of the evolution of the oceans from a warm world 50 million years ago to the cold one of the recent ice age is summarized in Table 1.

Table 1: Synopsis of the History of the Deep Circulation of the Ocean

Time	Event
More than 50 million years ago	Ocean can flow freely around the world at the equator. Rather uniform climate and warm ocean even near poles. Deep water in ocean much warmer than today. Only mountain glaciers on Antarctica.
35–40 million years ago	Equatorial seaway begins to close. Sharp cooling of surface and deep water in south. Antarctic glaciers reach the sea; glacial debris in sea. Seaway between Australia and Antarctica opens. Cooler bottom water flows north, flushes ocean. "Snow limit" drops sharply.
25–35 million years ago	Stable situation: partial circulation around Antarctica is possible; equatorial circulation interrupted between Mediterranean and Far East.
25 million years ago	Drake Passage between South America and Antarctica begins to open.
15 million years ago	Drake Passage open; circum-Antarctic current formed. Major sea ice around Antarctica which is glaciated. First major glaciation of Modern Ice Age. Antarctic bottom water forms; snow limit rises.
3–5 million years ago	Arctic glaciation begins.
2 million years ago	Ice Age overwhelms the Northern Hemisphere.

In this circuitous way we are beginning to tie together the various strands of information that can be obtained from the movement of the plates, the nature of the oceanic sediments, the fossils preserved in them, and a good deal of ingenious reasoning based on simple physical principles. The whole picture still looks rather like a carelessly stored old jigsaw puzzle just found on a forgotten shelf; many pieces are missing and the rest are discolored and sometimes damaged. At times this gives a nice Sherlock Holmesian flavor to our work, but it can also be quite frustrating if the pieces of the puzzle are too blurred or faded with age. New pieces are found every day, however, and we are looking forward to the time, perhaps not too far off, when we will be able to replace our inventive, but largely intuitive, explanations with rigorously computed models based on the laws that govern the behavior of the sea.

The Remote Past and the Near Future of the Snow Limit

How should we visualize the calcium carbonate cycle in the ancient oceans long before the start of the Antarctic glaciation and the present pattern of deep water circulation? There are several possibilities that could have occurred, but no certainties. First of all, decaying organic matter produces humic acids and carbon dioxide that can dissolve calcium carbonate. Organic matter does decay on the ocean floor today and contributes a little to the dissolution process, but the effect is trivial compared to the intense dissolution that results from the cold Antarctic bottom water heavily charged with carbon dioxide. On the other hand, in a more sluggish deep sea devoid of strong bottom currents, the decay of organic matter might be an important process (replacing the corrosive polar water) in the dissolution of carbonate. Earlier we discussed a variety of local sources of bottom water that might have played a role more than 50 million years ago. Some of these might have acquired corrosive properties while at the surface and might have flowed vigorously enough to dissolve sufficient carbonate to keep the cycle going. On the whole, however, one would expect that under such local control of bottom-water flow the snow limit would vary greatly from place to place.

Such a variation, however, is not observed. On the contrary, our data show that in the past the snow limit has been quite uniform worldwide, as it still is today, and that it has varied with time in a remarkably simple fashion. During the last 50 million years, changes in the depth of the snow limit have corresponded closely with changes in the extent of the Antarctic glaciation and consequently in the nature and flow rate of the bottom water. Prior to 40 million years ago, the snow limit was uniformly high in all oceans, indicating a very rapid recycling of calcium carbonate. The global uniformity of its position suggests that there was a global cause involved. It is possible that in those remote times the organisms of the sea surface were reproducing at a much faster rate than today and hence forced the recycling rate up to a very high level. Perhaps, but we have no independent evidence for such blossoming of oceanic life. On the other hand, there is a very striking similarity between the rise and fall of the snow limit over time and the flooding and emergence of the continents. When the sea flooded the land, the snow limit rose; when it receded, the snow limit fell. The explanation for this parallelism of sea level changes and carbonate dissolution might be as follows.

Users of carbonate are not restricted to the oceanic environment; on the contrary, they are abundant and very effective in warm and shallow seas. Furthermore, because water at shallow depths does not dissolve the skeletons, every ounce of carbonate extracted in a shallow sea is permanently lost. In the present world, this loss is not significant because the shallow seas are not extensive and are located in part in high latitudes where biological productivity is possible only during part of the year. During the flooding of 75 million years ago, however, the shallow seas not only were four times more extensive than today, but also were located mainly in warm climates. It is thus a fair possibility that life in shallow waters then extracted so much carbonate that the snow limit in the deep sea simply had to rise, by whatever means it could, to supply enough material for the demand at the surface. Shortly, the cores from the *Glomar Challenger* should provide enough information to calculate carbonate budgets for various times in the past and check this hypothesis.

The process of carbonate recycling in the deep sea has considerable practical importance. The corrosiveness of seawater for carbonate comes mainly from the carbon dioxide dissolved in it. This carbon dioxide, in turn, is acquired from the atmosphere in proportion to the amount present there. Human activity since the beginning of the Industrial Revolution has increased the carbon dioxide content of the atmosphere significantly by burning fossil fuels, such as coal and oil. The annual input of carbon dioxide is still increasing rapidly, and by the time not so far away when we have burned all of our resources of fossil fuels, we will have multiplied the carbon dioxide content of the atmosphere about four times. This in itself may have rather large consequences, because carbon dioxide absorbs a large part of the infrared radiation that comes from the sun and is reflected by the earth. If we increase the carbon dioxide content of the atmosphere, the surface of the earth will become warmer, perhaps much warmer. Dire consequences, including the melting of polar ice and the flooding of coastal areas, are foretold from this "greenhouse" effect, but nobody is sure yet what the magnitude will be. Will the increase in carbon dioxide also change the sea? About 40 percent of the added carbon dioxide will dissolve in seawater, where it will decrease the amount of carbonate and thus make the surface water less saturated. To keep the biological customers supplied, the snow limit must then rise and a considerable amount of the carbonate sediment on the seafloor will dissolve. Our knowledge is as yet a bit primitive, but the first calculations are reassuring; there seems to be plenty of calcareous sediment available to insure an adequate supply to maintain the standard of living at the surface.

The Origin of Seawater and its Constancy in Time

While we are on the subject of the chemistry of seawater, we should briefly consider its origin and the possibility of significant changes in its composition over time, a subject that we deferred earlier. The origin of seawater and that of the atmosphere are closely related, and there are two main possibilities. The first is that of a primitive planet endowed with a thick atmosphere, inherited from cosmic sources, consisting of methane, carbon dioxide, ammonia, and water vapor (somewhat like the planet Jupiter). The other possibility is that the water was a by-product of the gradual segregation of the primitive earth into an outer and an inner shell as the planet began to heat up. Volatile elements such as water and various gases were concentrated in the outer shell or released to the atmosphere. (Modern volcanoes still produce considerable steam from interior sources.) Gradual condensation of the vapors as the earth cooled then produced a global ocean. Neither of the two models has room for free oxygen in the first two or three billion years of the earth's history; that life-giving element was bound up in oxides from which it was only gradually released much later by the metabolic processes of life itself. The principal difference between the two points of view—and both probably contain elements of truth—is one of timing. Was the ocean fully present during the first few billion years of the earth's existence, or did it come into being slowly and is it perhaps still being added on to by steam from volcanic eruptions on land and under the sea?

Ample geological evidence exists to show that there was plenty of water around as early as three to three and a half billion years ago; we recognize water-laid sediments of that age as well as fossil algal colonies. The composition and volume of this water, however, may well have been quite different from today's. Seawater is a solution of many compounds, most of which come from the land, where they are leached by weathering and transported by rivers. Until recently it was assumed that leaching and weathering on the deep-sea floor, although possible, were so slow that they had little importance. The recognition that mid-ocean ridges are a zone of extensive volcanism has changed that.

Seawater can percolate in cracks and fissures of the young, hot rocks to considerable depth and leach certain elements from the rocks while leaving others behind. During dives in the research submarine *Alvin* on the

Mid-Atlantic Ridge, I have seen an abundance of such fissures, literally hundreds per mile, and have become convinced that they provide ample access for seawater to the hot rocks beneath the surface. Measurements and theory suggest that some elements, for example, calcium, silicon, and iron, may be extracted by circulating hot seawater in amounts not too different from those supplied by rivers. Others, such as potassium, magnesium, and sodium, are lost in similar quantities. The process is not restricted to the zone of volcanic activity itself; for many miles on either side, it appears to take place until the crust becomes too cool or the fissures are sealed with sediment.

The early earth probably had few continents, and elements extracted in this manner from the oceanic crust may have dominated the composition of the sea, which consequently may have been less salty but richer in iron and silicon than the present ocean. As the earth evolved, continents grew and the amount of material contributed to the sea by weathering on land and transport by rivers also increased. About one billion years ago, it seems that the present balance between the supply of elements from the land and from the ocean floor was approximately attained. Ever since that time, the composition of ocean water has probably varied only within narrow limits.

The very existence of oceanic life places rather strict limits on the composition, and sediments and mineral deposits confirm that the ancient oceans were much like ours. That does not mean, however, that there has not been some variation. Sediments and oceanic minerals are poor recorders of the conditions of their formation; they are unstable and are much modified during and after their deposition. The record they carry of the original environment is at best vague and ambiguous, and at worst, obliterated. We can easily conceive of several processes that might have varied the composition of seawater within the limits of tolerance of oceanic life. The release or uptake of chemicals by hot lavas is one. A rapidly moving plate would expose more hot rock and have a larger effect on the chemistry of the sea than a slow one, so that in periods of rapid movement and fragmented plates with many ocean ridges, the chemistry of the sea might have been measurably modified. The land as a provider of dissolved material is also subject to variation. A low continent, widely flooded by the sea and otherwise covered mainly by deserts, provides less material to the sea than continents with tall mountains, large rivers, and heavy rainfall. Since flooding of continents and fast plate movements might be related, the two effects could be cumulative.

Another possible cause of changes in the composition of ocean water is biological. In the present ocean, microscopic organisms are responsible for the large-scale recycling of carbonate we have just discussed. Other organisms use silica for skeletons with similar consequences for the chemistry of the ocean. Oceanic organisms have not always had the habit of using carbonate and silica for their shells with such abandon; the large-scale production of calcareous and siliceous skeletons in the open sea came into being rather abruptly about 150 to 200 million years ago. Before that time, although life forms in shallow seas built skeletons from these materials at least as far back as one billion years ago, there is no evidence that this process went on in the open ocean to any significant degree. When suddenly on a large scale it became fashionable to build skeletons out of carbonate and silica the change in the chemistry of the ocean must have been quite drastic. Why did this new biological practice take hold so suddenly? Was it a change in the composition of the seawater that permitted the construction of large quantities of calcareous and siliceous shells? Or was it the intrinsic evolution of life that finally equipped certain species to venture out into the blue water of the open sea and use the carbonate and silica they found there?

Although we have no answers for these important questions at the present, we know enough to sketch, in the next chapter, some of the possible relations between the evolution of life and the evolution of the ocean from the vantage point of the continental drift theory. If it seems that we discuss subjects more in terms of questions than in terms of answers, that is the way of science; the value of a new idea is measured either by the new avenues of inquiry it opens or by the misconceptions it removes. As Thomas Huxley once said, "Among public benefactors we reckon him who explodes old error as next in rank to him who discovers new truth." That is a consolation to those scientists, a vast majority I am afraid, who never discover new truth.

Etruscan fishing.

CHAPTER THREE

WATCH THE SWIRL OF LIFE IN THE WATER

O ye that stand upon the brink,
Whom I so near me through the chink
With wonder see: what faces there
Whose feet, whose bodies do you wear?
I my companions see
In you, another me.
They seemed others, but are we;
Our second selves those shadows be.
Thomas Traherne, 1637–74, *Shadows in the Water.*

THE SEA TEEMS with life. This fact is obvious to the most casual of wanderers across the tide pools of any coast. In mid-ocean, I have seen the deep-blue water turn yellow-green from dense masses of plankton, and once in a while one even has to stop the ship to clear thousands of salps—small, jellylike organisms—from the engine intake valves. The point was brought home most forcefully when, years ago, we were studying the continental shelf off southwest Africa and were surrounded at all times, on the gray and chilly sea, by blowing whales, frolicking herds of fur seals, and huge flocks of gannets, boobies, and albatross diving for fish. A primordial sight once common in many parts of the ocean, it has now become very rare.

This flowering of life, in large part microscopic, has major significance for us, because it furnishes one of our most important food resources. It is important also, in the context of this book, because marine life faithfully records the characteristics of ocean water masses and the oceanic circulation now and in the remote past. Thus their fossil remains allow us to decipher much of the history of the sea that otherwise would elude our understanding. Yet compared with the richness and variety of life on land, life at sea is sparse and is generally not highly diversified. About 200,000 different species inhabit the sea, only a small fraction of the number that live on land. Of these 200,000 perhaps 2 percent live near the surface in the open sea, a very small number on the deep-sea floor, and all the rest in the shallows and the intertidal zones of the coasts—mostly on the bottom in tropical waters. Compared with environments on land, those in the sea tend to be uncomplicated and are inhabited by rather simple communities consisting of only a handful of species, although each species may be prolific. Only in the intertidal zone—with its great variety of environments in the surf zone, between tides, in the transition between salt and fresh water, and on rocky or sandy shores—do the conditions and the responses of the inhabitants match the complexity of such common land environments as a temperate woods or a tropical rain forest.

All life on earth depends on food created by specialists out of inorganic materials, such as water, carbon dioxide, and oxygen, usually with the aid of sunlight. The organisms that perform this task are called primary producers. Because sunlight is rapidly absorbed in water, the zone where primary production is possible extends only 300 to 600 feet below the surface of the sea. Underneath, eternal darkness reigns. It is in the thin skin at the surface of the ocean that the base for all oceanic life is produced; anything living in the abyss below depends upon the rain of organic debris from above for sustenance.

There are fundamental differences between life on land and that in the sea. On land, the physical and chemical environment creates serious problems for survival; plants and animals must protect themselves against destructive or drying winds, droughts, erosion and sedimentation, the severe cold and lack of light of winter, and a whole host of other inanimate dangers. In the sea, by contrast, the environment, with the exception of the near-shore zone, is generally friendly to life; organisms have to compete only with each other. But because the environment itself, especially in the open sea, affords virtually no protection against predators, this competition is no small problem. The threat from other life, however, demands strategies of defense different from those required by the threat from the elements.

Thus whereas life on land must invest considerable energy in structural materials such as bone and wood, this is not required for organisms suspended in the sea. They, on the other hand, must counteract predation by rapid reproduction, fast growth, or high mobility. Even if they adopt mobility as a means to survive, moving about in the sea requires far less energy than on land or in the air, where the force of gravity must be counteracted. Consequently the predominant organic material on land is carbohydrate, whereas in the sea it is protein. Carbohydrates store energy very effectively for structural strength or for times when the food supply is low, during winter for instance, but organisms whose metabolism is mainly based on carbohydrates are slow to grow and multiply. Protein, by contrast, allows a quick energy release, although it does not store energy as effectively; a protein-based economy makes possible rapid growth and reproduction and excellent locomotion. In an environment where predators are the principal threat, the ability to reproduce rapidly and create a quick succession of individuals that are small in size but large in number is an excellent strategy for survival. It has been widely adopted by the plants of the sea (the primary producers) and by the animals that graze on them. Another practical solution is rapid growth to a size larger than the predators can handle, a strategy followed commonly by the higher level predators at the far end of the food chain.

Food From the Sea

The fertility of the sea, its ability to support life, is reflected in the total amount of organic matter present. In terms of the main constituent, organic carbon, the average annual production in a coastal zone is about four ounces per square yard of area, whereas in even the most fertile parts of the open sea it is only one ounce. Averaged for the whole ocean, the amount of organic carbon produced annually per square yard is about two ounces. In terms of plant material or fresh fish, one ounce of organic carbon represents several ounces of organic tissue, but even the largest production values in the sea are very small compared to the production of organic matter of a tropical jungle or a well-managed Iowa farm.

The enormous variation of the productivity of the sea from region to region is not always fully appreciated when the food potential of the ocean is discussed. High productivity values (four ounces or more of organic carbon per square yard of area per year) are restricted to a small part of the sea surface—some 10 percent—located mainly near-shore. The rest of the ocean area, which is about 70 percent of the surface of the earth, varies from being not very productive to being a veritable desert by standards of productivity on land. Broadly, we can divide the productivity of the ocean into three classes: the open ocean, often called the *pelagic* region; the coastal zone; and regions of so-called *upwelling* that are situated along the west coasts of continents.

The phenomenon of upwelling results from the common occurrence of equatorward winds along west coasts. Under the influence of the Coriolis force, these winds and the marine currents they generate are deflected to the right on the Northern Hemisphere and to the left on the Southern Hemisphere. In each case the result is an offshore drift of water that must naturally be replaced. This is accomplished by water welling up from below. The upwelling water is cold and comes from dark depths below the zone where sunlight permits primary production; it is therefore rich in nutrients (since they have not been depleted by plant growth) and is very fertile. The upwelling thus gives rise to enormous plankton blooms, often so dense that the sea takes on the look of diluted pea soup. These blooms in turn feed large quantities of fish. Upwelling is normally a seasonal phenomenon. On the California coast, for instance, upwelling occurs in the spring and early summer when, because the upwelling water is so much colder than the air over the adjacent land, it produces the extensive coastal fogs and low clouds so disappointing to tourists. Many of the world's largest fisheries, such as those off Peru and southwestern Africa, are sustained by upwelling, as were the former sardine fisheries of California.

Estimates of the total primary production per year of organic matter in the oceans and the quantity of fish that this production can support vary considerably. Table 2, after John Ryther, a marine biologist at Woods Hole Oceanographic Institution, is representative.

Table 2: Primary Organic Productivity of the Oceans and the Quantity of Fish It Can Support (after John Ryther, *Science*, v. 166, 1969, pp. 73-74).

	Primary Production in Tons of Organic Carbon Per Year	Percent	Total Available Fish in Tons of Fresh Fish Per Year*	Percent
Oceanic	16.3 billion	81.5	0.16 million	0.07
Coastal Seas	3.6 billion	18.0	120.00 million	49.97
Upwelling Areas	0.1 billion	0.5	120.00 million	49.97
Total	20.0 billion		240.16 million	

*The numbers in the third column are available fish, not fish obtained by man.

The main thing that is noteworthy in this table is that the open ocean, even though it represents roughly 90 percent of the total sea surface, produces a proportionally smaller percentage of the total organic matter and only a minute fraction of the total quantity of available fish. It is further obvious that coastal fish are much more efficient in utilizing their food than are pelagic fish, and the fish in upwelling areas are real marvels. This phenomenon —the efficiency of use of the organic carbon resulting from primary production—is a key element in the potential of the sea as a food source; it requires some explanation.

On land, we humans eat the cow that eats the grass. Although a cow is not a very efficient converter of grass into meat—nor does grass convert sunlight and carbon dioxide into organic matter with any great efficiency—the fact that we, the human consumers, are only two steps removed from the grass insures a minimum loss of primary food. With marine food sources, such as tuna or lobster, the situation is quite different. Here we generally do not eat the herbivore that consumes the oceanic equivalent of grass, because this herbivore is microscopic in size, lacks any particularly appealing taste, and is thinly dispersed and hence hard to catch. Instead we wait for these herbivores to be eaten by a larger animal, often a small crustacean, for it in turn to be consumed by a higher level carnivore, perhaps a small fish, and so on, until we arrive near the top of the food chain and consume the herring or the tuna. The equivalent on land would be for us to eat lion or wolf meat rather than beef, making us dependent on a very small number of wolves rather than on great herds of cattle.

On land and at sea, every additional step before we hook into the food chain reduces the amount of organic matter that we can harvest as food. In the open ocean, where the number of steps between the microscopic plants and the fish we eat is about five, the ultimate efficiency with which we recover the primary production of organic matter is about 10 percent. In upwelling areas, where the excellent nutrient supply leads to very large and numerous primary producers, the fish themselves have learned to become, in part, grazers; when we eat these fish we are only one or two steps up the food chain, with an efficiency of about 20 percent. The coastal zone occupies an intermediate position between the open ocean and the upwelling areas in terms of the efficiency with which man can utilize the primary production of organic matter. Table 2 shows this variation in the efficiency of recovery of the initial food quite clearly. In case one wonders what happens to the food lost at each step along the food chain, it is used for energy in locomotion, for metabolic functions, or becomes waste.

Can We Increase the World's Fish Supply?

Our knowledge has not yet advanced enough to determine very precisely what the productivity of the ocean is, although the numbers in Table 2 are probably not far off. Nor do we know for sure what these numbers mean in terms of maximum harvestable fish. At present most experts think that something like 100 to 150 million tons of fish per year is a reasonable estimate of a sustainable harvest that will not in the long run deplete the fish stocks. Probably little would be added if we were to discover new stocks or learn to use those that are not currently fished. The world catch of marine fish has risen rapidly from 40 million tons in 1960 to 70 million in 1974 and remained at about that level in 1975. This average rate of increase, if maintained, will bring us to the level of maximum sustainable yield, or even exceed it, within the next decade. This implies that, as a potential resource to feed the hungry of the world, the sea cannot be expected to provide much additional help. Not all experts are so pessimistic, but the optimists are decidedly in the minority. In addition it is clear that by far the best fishing grounds are in the coastal and upwelling regions; hence anything that damages the ecology there, such as oil spills, offshore construction, or pollution from sewage, will have a disproportionate negative effect on the marine food supply.

Are the widespread hopes, so often expressed in the news media, that large segments of the world can be fed from the sea just an illusion? So it would seem, but such a conclusion is intuitively dissatisfying. We have heard a great deal of doomsday talk in recent years, but much of it has proved to be wrong. Although man is fully capable of doing himself in, I tend, looking at history, to feel that human ingenuity may yet be capable of solving most of the problems that threaten us. Fortunately, there are several ways in which one might circumvent the limits that the sea seems to have placed upon our use of its living resources.

For instance, we might consider harvesting lower links in the food chain rather than the highest ones as we do now. We would thereby avoid the sharp drop in efficiency that occurs at each step when the higher level exploits the lower one. If the herring converts 10 percent of its food into flesh that we can eat, then, by directly harvesting its food rather than the herring itself, we should be able to increase our return, although not tenfold, yet perhaps threefold to fivefold. Of course, in doing so, we would directly compete with the herring and deprive it of much of the food it consumes. Fewer herring would survive, and we could no longer count on a good herring catch. The basic problem with harvesting lower links in the food chain, however, is the same one that causes the herring to be so inefficient in converting its food into flesh. In order to catch its food, which is small and appears in abundance only seasonally, the herring must chase around a great deal, cover a large territory, and survive for long months on the food stored in its own tissues. All of this consumes most of the energy gained by eating the catch.

If we harvest fish high up in the food chain, we take advantage of their particular strategy for survival, namely the ability to grow large quickly, a device that makes it easy for us to harvest the entire primary food production of a sizable piece of ocean over a considerable time. Catching a single whale, a reasonable undertaking in terms of human time and energy, provides us with the aggregate output of a large portion of polar sea over many years. If we do not hunt the whale, in order to obtain the same quantity of food for ourselves we must filter an enormous volume of ocean water repeatedly over ten years or more. Similarly, it would not be enough to catch all the plankton in a given area at a given time to equal the food we can obtain from the herring of that region, because herring grow over several seasons and thus harvest the plankton output several times.

Attempts to harvest lower links in the food chain clearly would be difficult and costly. In fact, unless wholly new fishing techniques are devised that we cannot even envisage, the efficiency gained by going lower in the food chain would be offset by losses in time, increased use of fuel, more fishing vessels, higher costs of processing the catch, and many other problems. Moreover, few people would enjoy eating plankton directly; I understand from those who have tried that it has the consistency of flour and is quite tasteless. If we convert the plankton into something more palatable, for instance by feeding it to chickens, we would again lose a significant amount of efficiency. Nevertheless, some possibilities do exist. The great Antarctic whale herds of the past fed on a fairly large shrimp-like creature, the krill, which lives in the cold and fertile southern polar waters in great abundance and is large enough to be caught economically. Now that the whale herds have largely disappeared and those remaining are no longer a match for the krill output, krill fisheries are being contemplated by the Soviet Union, but what to do with the product remains unsolved.

We can, however, approach the problem from a different perspective. Today we harvest the sea not as farmers but as hunters, taking the crop we want and relying on nature to restore the balance. But this is not the sole course open to us; we can ranch or farm the sea in a variety of ways, improving the yield of a species we wish to harvest. Aquaculture has been moderately successful, albeit on a small scale, in enhancing catches by protecting a desirable fish, such as the salmon, during part or all of its life cycle. It is fed, protected against disease and competition, and watched over to minimize the chance of its not returning to the place of the catch. Undoubtedly, this approach will be used more and more widely in the future, but it is costly and is unlikely to increase the global fish production significantly for quite some time. If it became a widespread practice, we would need to think about what our selection of a few favored species might do to the balance of nature in the ocean.

Another very interesting possibility is to increase the fertility of the sea by enhancing naturally occurring processes, such as that most successful experiment in marine fertilization—the upwelling process. The primary food source of the ocean, the microscopic oceanic plants, drift freely with the currents, although they have some small-scale mobility which allows them to protect themselves from sinking below the zone of sunlight or to move from a depleted drop of water to one richer in nutrients. Because of this essentially passive mode of living, growth and reproduction come to a quick halt once the nutrients in the immediate surroundings have been fully used up.

Sunlight of course is in unlimited supply, and oxygen and carbon dioxide are replenished continuously from the vast reservoir in the atmosphere, but small amounts of other substances, such as nitrogen, phosphorus, copper, and some organics (e.g., vitamin B12), are also required to sustain life. All of these substances are dissolved in ocean water in small amounts, but a strong bloom of plankton can exhaust the supply in a matter of days. In the depths of the ocean, however, these nutrients are present in abundance because they are not utilized there and are constantly replenished from decaying organic matter that sinks from above. If there were a means to bring the deep water to the surface, this reservoir could keep the primary production process going nearly indefinitely. This, however, cannot be achieved easily. Because deep water is colder and thus denser than surface water, it will not automatically rise. Nature uses a variety of devices to bring deep water to the surface, such as upwelling caused by offshore winds or the stirring caused by the strong currents of the equatorial zone or by large ocean waves. In the polar seas seasonal cooling of the surface water to a very low temperature also causes the surface water to become denser and sink and the deeper water to rise.

Oceanographers, such as Sam Gerard of the Lamont-Doherty Geological Observatory at Columbia University, have been pondering for years how one could give these natural processes an artificial push and thus cause the ocean to fertilize itself more efficiently. The best way would be to heat the deep water several hundred feet below the zone of sunlight to make it less dense and thereby cause it to rise. Although the temperature difference required is not very large, water is notorious for the amount of heat it takes to warm it even one degree. Furthermore, in order to have a measurable effect, because the population densities are so small the area over which the vertical circulation must be enhanced needs to be very large, probably hundreds of miles in diameter. Potential sources of leftover energy that could be tapped to provide the needed heat include cooling water from a coastal or possible future offshore nuclear power plant, or even radioactive waste from nuclear installations—carefully sealed to prevent the radioactivity (but not the heat) from escaping. If the heat source were placed at the proper location and depth, one might ultimately gain considerable benefit from a now-useless by-product. Gerard is conducting a small-scale pilot project in the Virgin Islands, using wastewater from a power plant. The project appears to be promising, but it is too early to tell whether the technique is capable of changing the ocean's fertility on the truly gigantic scale demanded by the food shortages of the future world.

Oceanic Circulation Mirrored in Marine Life

The biological communities of the sea, although simpler than those on land, correspond as closely to variations in the oceanic environment as, for instance, the plant life does on land. Salinity variations, the large temperature gradient from the equator to the poles, seasonal changes in the conditions at the sea surface, the differences in fertility between the rich waters of the major current systems and the oceanic deserts of the central gyres are all reflected in the distinctive character of their inhabitants. The closest correlation between environment and life exists in the plankton, because it cannot move over large distances and must stay with its own water mass. When looking at distribution maps of organisms in the open ocean, we observe a striking zonality parallel to the latitudes, with dramatic changes in the composition of the plankton as we move from the equator to the north and south. The coastal currents, upwelling areas, equatorial currents, central gyres, and subpolar and polar regions each have their own, highly characteristic plankton communities. Furthermore, we can distinguish fertile waters from infertile ones by the amount of organic matter that is being produced.

The majority of the organisms that live near the surface never make it to the bottom after death. They are eaten and their predators are eaten again; the organic debris decays, releasing its nutrients into the water; and only a very tiny fraction of what is present in the surface layer eventually reaches the bottom where it is further diminished by bottom-dwelling animals. Shells and skeletons made out of hard materials, such as calcium carbonate and silica, best survive the trip and become part of the sediments. Their remains are often quite characteristic of the surface water mass in which they once lived. In fact, they can be so typical of their environment of origin that geologist John Imbrie of Brown University and his co-workers have been able to devise mathematical equations which relate the fossil assemblages precisely to the temperature, salinity, and fertility of the surface water mass from which they came. Because different organisms thrive in different seasons, these equations can even tell us the difference between the summer and winter water temperatures of the sea surface above the deposit of sediment. We can also safely assume that at least for the past several hundred thousand years or so the relation of the fossil assemblages to the properties of their home water masses has not changed significantly. Consequently the use of these equations has permitted Imbrie and his team to construct such astonishing things as maps of the surface temperatures of the oceans in mid-summer and mid-winter 18,000 years ago (*Science,* vol. 191, 1976, p. 1132). Such maps are of great value in reconstructing ancient climates, as we shall see in the next chapter. (A good example is the North Atlantic during the last glacial [see Figures 6 and 6a].)

If we go back further in time than a few hundred thousand years, the reconstruction of the ancient oceans becomes progressively more difficult, because we encounter more and more fossils of species that are now extinct. For those of course we are not nearly so sure what kind of environment they were adapted to. We can still draw some conclusions, although they become progressively less precise, as far back as 100 million years ago, but

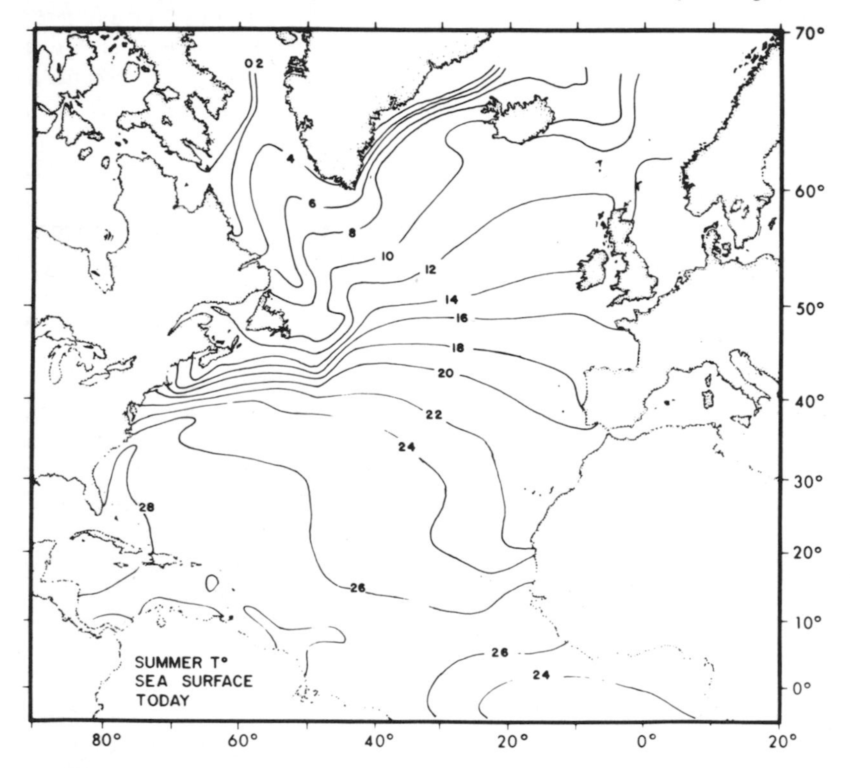

Figure 6: The temperature distribution in the North Atlantic Ocean during the summer at the present time and 18,000 years ago (opposite page) during the height of the last glacial period. The temperature distribution 18,000 years ago was obtained by analysis of fossil assemblages in sediments of that age. Note the very strong gradient of temperature across the Atlantic at the latitude of Spain that separated a warm southern area from the arctic northern North Atlantic in glacial time. It contrasts markedly with the present gradual northward temperature change. The temperatures are in degrees centigrade; continental ice caps (with hachured borders) and pack ice (with granulated borders) are indicated for the glacial period on the map in Figure 6a. The figures are from an article by A. McIntyre and N.G. Kipp in *Memoir 145* of the Geological Society of America (1976, pp. 60-61).

we are forced to use other and less precise information and must engage in entertaining but risky circumstantial reasoning.

Nature has been kind to us paleo-oceanographers by providing two major categories of primary producers and two major low-level predators equipped with skeletons or shells that can be preserved in the ocean's sediments, because they consist of calcium carbonate and opal. After giving us this much access to the past, however, nature has hastily set up roadblocks to man's curiosity which can be very effective. The principal obstacle is the dissolution of both carbonate and silica as they sink through the water and settle on the bottom. We have already discussed the dissolution of calcium carbonate in Chapter Two and have seen how the amount that finally becomes part of the sediment depends on the amount produced at the surface, the variation of the dissolution with depth, and the depth of seafloor itself. Since some species have carbonate skeletons that are very susceptible to dissolution whereas the skeletons of other species are quite

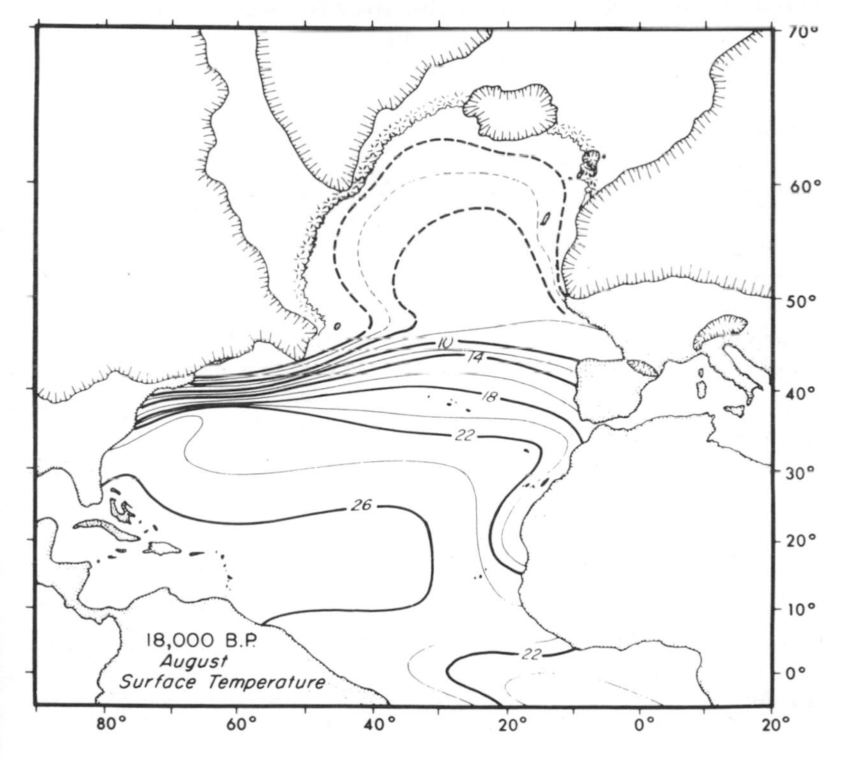

Figure 6a: Temperature distribution in the North Atlantic Ocean 18,000 years ago.

resistant, the assemblage that we sample on the ocean floor differs considerably from that originally living at the surface, and varies from place to place and time to time with the ocean depth, the fertility, and the changes in time of the dissolution process.

Siliceous skeletons are also strongly dissolved, but the amount of dissolution is not a function of depth as it is with calcium carbonate; hence we do not have to take account of depth changes of the seafloor as we try to interpret the meaning of siliceous fossil assemblages. Seawater dissolves silica at all depths, even at the surface, and diatoms and radiolaria can make their siliceous skeletons only because they protect them with a coating of organic matter. When this coating decays, the silica begins to dissolve, and only the most robust skeletons reach the bottom. The dissolution of silica is thus even more severe than that of calcium carbonate—less than 2 percent of what was formed at the surface becomes part of the sediment. Although even this small residue can tell us much about the environment at the surface, it is distinctly uncomfortable to draw important inferences from so small a remainder of the original community.

The bulk of all oceanic sediments consists of calcareous and siliceous skeletons and shells. It is one of the ironies of nature that, notwithstanding the chemical simplicity of the main components, it is excruciatingly difficult to determine their proportions precisely. Carbonate of course is easily analyzed quantitatively. Not so silica. A decade ago, my students and I spent a great many months in the lab trying to determine accurately the amounts of silica present in deep-sea sediments. We looked through microscopes; we tried to float the lighter shells out of the sediment with foam or in liquids of precisely controlled density; we tried chemical analysis, X-ray techniques, infrared spectroscopy—a whole arsenal of approaches—and none really worked. Our successors have not been much luckier, and even now the determination of silica in ocean sediments is tricky, time-consuming, and much less precise than we would like.

Nonetheless, there is often enough information in the skeletons that have escaped dissolution that we occasionally succeed in drawing remarkably fine conclusions regarding the ocean environments of the past. For more than eight years, the deep-sea drilling ship *Glomar Challenger,* financed by the National Science Foundation, has been taking cores of sediments in the deep ocean that date from the present to 150 million years ago. From these voyages has come a core library that contains a rich and wonderful record of the history of the oceans and of the evolution of oceanic life. The record is not easy to read, and those of us who are engaged in this endeavor are often intensely frustrated by our inability to understand what nature is trying to tell us. Piece by piece, however, we are devising the necessary techniques and approaches and are unraveling the information which will be the building material for new theories, sometimes stumbling on important new insights purely by accident. In the next decade we should advance a long way toward achieving a complete reconstruction of the history of the oceans, climates, and marine life over the last 150 million years or more. Perhaps, if we are fortunate, we may even come a little closer to answering the weighty question of how much the evolution of life has been a matter of heredity and to what extent it has been a response to a changing environment. I see the coming decade as another challenging and exciting period in the earth sciences, similar to the continental drift revolution we have just behind us, but promising new and possibly deeper insights about life itself. It is much too early to define the ideas that may evolve, but the following speculative considerations give a foretaste of their possible flavor.

Catastrophic Changes in Marine Life

For the past several years I have spent much time studying the history of the oceanic circulation as it is recorded in the sediments and fossils of the oceans. While turning the pages of these chronicles of the earth, I have become impressed with the drastic changes in the nature and composition of marine life that have occurred from time to time during the last several hundred million years, short as this time span may be in relation to the three billion years that life has existed on earth. I note here that the history of the earth is very long and our ability to resolve events that followed each other closely in time is very limited; consequently an event that took something like a million years to run its course will appear to us to have been instantaneous, and if the event produced a large change, geologists tend to describe it as catastrophic. Even in the most recent part of the history of the earth, however, such a catastrophe may represent the developments of tens of thousands of years, and in the distant past—half a billion years ago for instance—a catastrophic event may have lasted for millions of years.

About ten years ago, Milton Bramlette, a geologist at Scripps, wrote an intriguing paper (*Science,* vol. 148, 1965, p. 1696) in which he pointed out that 65 million years ago, at the boundary between two major eras of geologic history, the Mesozoic and the Cenozoic, life in the oceans changed dramatically as a result of a massive extinction of earlier life forms and an explosive evolution of new ones. On land the disappearance of the dinosaurs opened the way for the takeover by the mammals, but the transition was gradual and totally unlike the very abrupt transformation of oceanic life. Bramlette suggested that at this time, when the continents were low and flooded to a maximum extent, the supply of nutrients to the ocean would have been drastically curtailed because they had been siphoned off

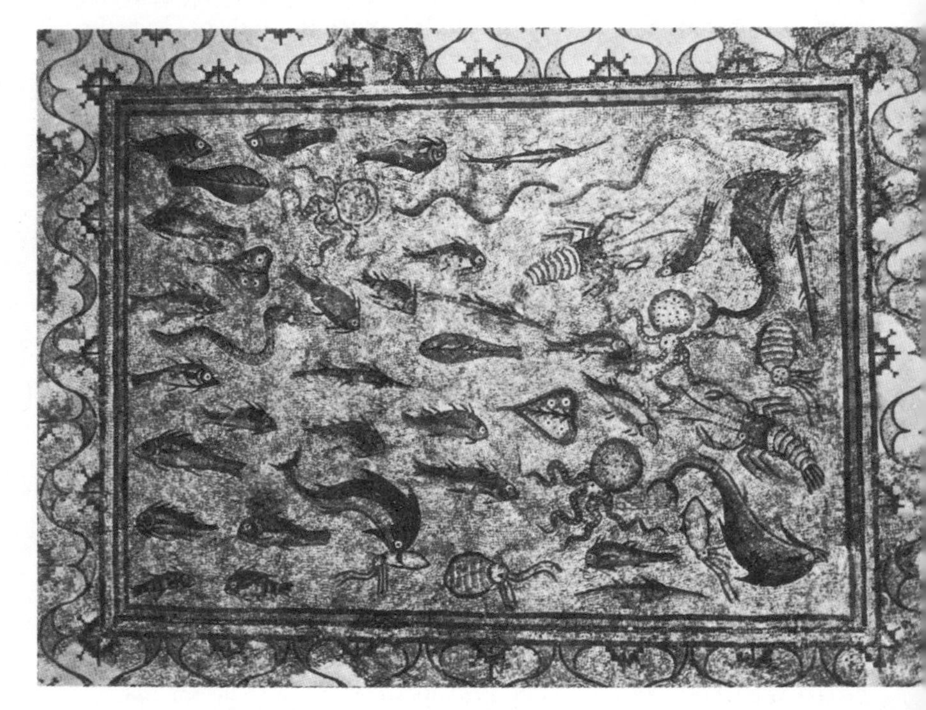

Roman mosaic.

by the inhabitants of the shallow seas. As a result, the open-sea flora and fauna, deprived of their food supply, died in masses. Only the most hardy and tolerant survived and provided the base from which numerous new species gradually evolved to occupy the available space when the seas withdrew again from the continents and the nutrient supply was restored.

This catastrophe at the boundary of the Mesozoic and Cenozoic is not the only drastic change in the composition of marine life that we know about; in fact, it is not even the most striking one. About 250 million years ago, for instance, at the time of the supercontinent Pangaea, the known number of fossil marine species dropped by nearly 60 percent—a change in the composition of the marine flora and fauna that has not been equaled since that time. In order to describe these changes in the makeup of life, paleontologists use a concept called *diversity*, which characterizes with a single number the variety of species present in any given group, in any chosen region, or during any specific geological period. In its simplest form diversity is merely the total number of species (or genera, families, orders) that can be found in the fossil record in such a given category. It is a simple notion and can be severely criticized on several grounds. For example, in a sample consisting of several thousand individuals of one species, one additional individual of a second species increases the diversity to two. It would also be two if the sample consisted of 500 individuals of each of two separate species. Obviously the implications of the two cases are quite different. In the fossil record the total number of species seen is clearly a minimum; many organisms were never fossilized because they did not have hard skeletons, lived in places where they could not easily be preserved, or have not yet been discovered. Thus the diversity of the present life on earth is many times greater than that of any past period.

Notwithstanding the flaws in this concept, the changes in diversity with time are striking and probably meaningful. They show, as paleobiologist James Valentine of the University of California at Davis has pointed out, some remarkable parallels with the history of continental drift (see Figure 7). During times when the continents were joined together, the diversity tended to be low, but it increased markedly as soon as a supercontinent split up and the pieces began to drift apart. One can argue that we have really only two examples of a supercontinent and subsequent drift in the geological record and that our knowledge of the older sequence of events and its fossil record is very deficient. I cannot but grant that point, but it is illuminating, nevertheless, to speculate on the possible causes of the correlation between continental drift and fossil diversity. In doing so, I follow James Valentine who summarized his views on the subject in an article in *Nature* (vol. 228, 1970).

One can reasonably assume that the number of different types of orga-

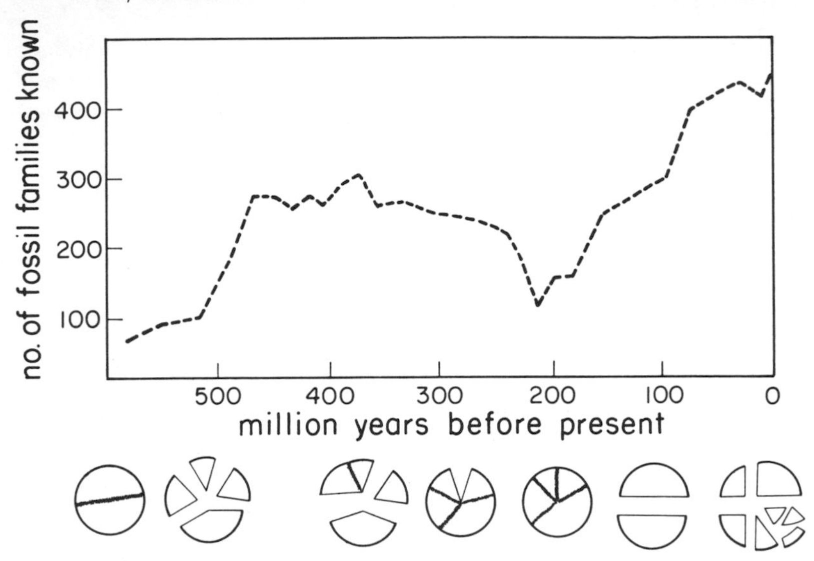

Figure 7: Changes in diversity over geologic time that seem to reflect the drifting of the continents. The dashed line depicts the variation over time of the number of families of shallow marine organisms. The jigsaw puzzles at the bottom illustrate (from left to right) the shifting of the continents from a supercontinent about one billion years ago to dispersal, followed by another supercontinent 200 million years ago (Pangaea), and another phase of dispersal. This diagram is modified from one by James Valentine in *Nature* (vol. 228, 1970, p. 657).

nisms that make up a community is strongly determined by the supply of food. The food supply, in turn, must be considered from two aspects: the *level* of supply, namely whether or not food is readily available and abundant, and the *stability* of the supply, whether or not the organisms can count on the same supply at all times or must reckon with times of feast and times of famine. If the supply of food is large, almost any type of organism can exist in large numbers, efficiency is not very important, and the population can rely on sheer numbers for survival rather than on special adaptations. On the other hand, if the food supply is small, a premium is placed on its efficient use, the environment will not support large numbers, and specialists that can make maximum use of the available resources have a great advantage over unspecialized organisms.

Unstable food sources require different strategies. In times of famine there will inevitably be mass mortality; it is therefore useful to build up large populations in times of abundance so that some may survive periods of future austerity. It is also advantageous to be tolerant of changes in conditions and able to use more than one type of food. When the food supply increases, organisms that are able to multiply rapidly can take advantage of the abundance better than those that reproduce more slowly. If the organisms are also flexible in their choice of a place to live, they have a better chance to hunt out diminishing resources when the food supply drops again. By contrast, with a stable supply, whether large or small, there will be no mass mortalities and thus no advantage in large numbers. The individuals will need only a narrow tolerance of environmental conditions because, once established and adapted, they can count on their share of the resources. Fast breeding is unnecessary, but specialization is important.

Genetically, a highly specialized organism has less potential to evolve to fit a new set of conditions than a less efficient but more tolerant one. Furthermore, if the number of individuals of a species is large, there is also a greater chance that this large gene pool contains the potential for evolving new and better adapted forms. Thus in an environment where the food supply is small but stable, specialization is a necessity to take advantage of the last crumb in the oddest spot, and a high diversity results. At the opposite end of the spectrum, a rich but unstable environment will at times support a very large population, but requires tolerant species that can multiply rapidly and adapt well to change. The diversity is low, but the genetic potential for future evolution is high.

Prior to the catastrophe that befell marine life 65 million years ago, the food supply in the warm and simple ocean was probably quite stable. Food supplies may or may not have been abundant, we do not know, but in either case this was a good setting for a high diversity, possibly with relatively small populations. Bramlette believes that the food supply diminished drastically 65 million years ago, resulting in large extinctions.

Valentine argues, however, that such a drop in a stable food supply would simply increase the amount of specialization; more specialists in ever smaller numbers would evolve to take advantage of every last bit of food in every possible locality, and the diversity would increase, not decrease. On the other hand, if the environmental change was not a drop in the *abundance* of the food supply—it might even have been an increase—but a decrease in the *stability* of the supply, one would expect the resident specialists, adapted to their dependable, small environments, to die off. Because the specialists constituted a majority of the species in the previously stable ocean, their extinction drastically reduced the diversity. Not affected by the change in environmental conditions, or at least much less so, were those species which, being broadly tolerant and prolific, could survive large temporary mortalities. Such species were small in number, but they contained the large, adaptable gene pool necessary to produce the distinct but gradual evolution (and increase in diversity) that followed shortly after the catastrophe.

Meditation by the Sea by an unknown artist.

Is such a change toward a more unstable food supply probable? I think it is. The catastrophe of 65 million years ago coincides with a time in which the continents had completely broken up and had acquired very long shorelines. They had become dispersed toward higher latitudes than before, thus increasing the climatic range, and the climate itself had developed a more extreme variation from the poles to the equator. The continents, previously flooded, were emerging, thereby reducing the areas of shallow sea and placing the rivers much closer to the deep ocean. The coasts were much closer to the edges of the continental shelves and therefore closer to deep water; this enhanced upwelling and provided an abundant but seasonally unstable food supply.

The same reasoning that was just applied to the catastrophe of 65 million years ago holds equally well if we look at the change from a single supercontinent to a dispersed situation about 150 to 200 million years ago. Again, stability was replaced by instability, causing first a reduction in diversity when the specialists became extinct, followed by a gradual increase in diversity as the remaining tolerant species began an evolutionary adaptation to the new environment.

If we go back even further to a supercontinent that may have existed perhaps a billion years ago and examine the effects on life of its breakup and dispersal, there is an additional factor that comes into play. So early in the history of life, many of the basic forms had not yet evolved, and many environments on earth had not yet been inhabited by life forms suited to their conditions—no plants on land, perhaps no organisms to take advantage of the opportunities of the open sea. Thus the burst of evolutionary expansion that followed the breakup 500 million years ago of this early supercontinent may have derived as much or more from the stimulus provided by the occupation of new territories and new environments as from the simple factors of stability and level of food supply.

If we look rather critically at the current state of our knowledge of this fascinating subject, we have to admit that it is not good. There are tantalizing glimpses of possible causal relations, of explosive evolutions and dramatic extinctions, of the interaction between the environment and the creation of new species, of a role for the dynamics of the earth in the dynamics of life. But glimpses are all they are, and many of the learned essays I regularly open with great expectation turn out to be fascinating concepts without much basis in fact. Sometimes, after scraping away the academic jargon or the mathematics of computer models, there are not even any fascinating ideas underneath. The remote past raises the most fundamental and intriguing questions about the origin and evolution of life and the evolution of the physical environment of the earth, but the answers may be some time in coming. Fortunately, the recent past is more accessible, and the questions it raises about the evolution of oceans and climate and the evolution of man are important and are more easily approached. The impact of these questions transcends the past and reaches into the future. With anticipation of vistas yet to unfold, we continue to watch the swirl of life in the water.

The Hunters in the Snow, detail, by Pieter Brueghel.

CHAPTER FOUR

IN GRANDFATHER'S TIME YOU COULD SKATE EVERY WINTER

I do think that, of all the silly, irritating tomfoolishness by which we are plagued, this weather-forecast fraud is one of the most aggravating.

Jerome K. Jerome, *Three Men In A Boat,* 1889

WHEN I WAS GROWING UP in Holland I used to view with a great deal of skepticism the old Dutch paintings of winter landscapes: thatched farmhouses covered with snow huddling in a bare, wide, and flat country with skeleton trees as a backdrop for innumerable gay and colorful skaters on the frozen canals. Our own winters tended to be wet, windy, chilly, and generally dismal; if we could skate at all, it was a rare event, possible only on some flooded meadow. For centuries, my family has resided with great persistence in the Dutch river country, but my father's stories about solidly frozen rivers that served as roads for months on end seemed to me a faulty perspective on the past rather than a historical reality.

Most people think of the climate as basically constant, even though we know that the weather is quite variable. We recognize changes in weather from year to year, even quite large ones, but assume intuitively that the average weather over the years—which means the climate—is today much as it has been for hundreds if not thousands of years. We all know that there once was an ice age when most of the northeastern and midwestern United States was covered with ice thousands of feet thick, but it was so long ago that we regard the occurrence as of only geological and perhaps anthropological interest. Most of us realize that the ice might someday return, but we are imbued with the spirit of the old geologists who believed that all geological processes were infinitely slow and that every major change took an infinitely long time. And so people assume that, if an ice age returns at all, it will be so slow in coming that no single generation will be aware of it.

Actually, these assumptions are false. New techniques have enabled us to date the ice ages much more precisely than in the past, and we have discovered that these catastrophic events are anything but slow in coming and going. For example, a short glacial interval that occurred between 10,000 and 11,000 years ago reached its peak in less than a century and disappeared equally fast. For several hundred years it replaced the temperate forests of England, western Germany, and the Low Countries—which then had a climate and vegetation much like that of southern Sweden today—with tundras, howling winds, and drifting snow. Imagine the consequences of such a drastic and rapid change in climate today in overcrowded northwestern Europe.

Geologists have assembled a great deal of information that tells us not to think of an ice age as a series of brief glacials of unremitting cold separated by tens of thousands of years of pleasant interglacials with climates much like the present. During the past two million years, the climate most of the time has been glacial, with fluctuating but generally severe arctic or subarctic weather in all northern and southern temperate zones. The interglacials, although they occurred regularly, were much shorter than the glacials and, on the average, not nearly as comfortable as the climate to which we are accustomed.

In fact, the weather of the last half century or so has been quite exceptional and represents by far the mildest and most favorable period for the last 1,000 years. To find a climate that was noticeably better than at present we must go back to the period from 5,000 to 7,000 years ago. Our present climate is typical of perhaps 3 percent or less of the past 100,000 years; the rest of the time the world was dominated by harsher conditions. On this basis one can logically argue that the weather in future years is likely to deteriorate and that the only question is *when*. Even the timing of the inevitable deterioration is not too much of a mystery, because interglacials last about 10,000 years on the average, and ours has already lasted that long. Thus we seem to be due for a change. Of course, the human time scale is rather short compared to those millennia, and the inevitable change may still lie a few generations in the future.

That another glacial period might not arrive for another few centuries or so is not as much of a comfort as one might think. Although the transition from an interglacial to the peak of a glacial takes a long time, we now know that climatic variations of shorter duration and less severity (but nevertheless significant in human terms) are superimposed on the glacial-interglacial-glacial sequence. Transitions into and out of a "mini-glacial" take place over much shorter time spans, on the more human scale of 50 to 100 years. In the past, such short-term climatic fluctuations have led to famine, war, and economic disaster in small, subsistence-level economies, but seldom had global consequences because the societies of the past were rather simple and self-sufficient. In our present complex, closely interconnected, overpopulated, and technologically advanced world, the impact of such climatic fluctuations would be quite different.

The winter of 1976-77 was unusually severe—the worst in a century—and resulted in serious shortages of natural gas in many parts of the United States, significant crop damage, and other negative effects that reverberated through the economy. Suppose that we knew for sure now that the year 2000 A.D. would bring much longer and colder winters and cool, short summers in the United States and Canada. The development of new energy sources for heating, the hybridizing of new wheat stocks adapted to a shorter growing season and lower temperatures, and the manifold other adjustments of an interdependent society would require huge capital investments and decades of planning and implementation. Of course such a climatic change would be unlikely to affect only ourselves, and the failed harvests elsewhere, along with a global increase in energy demands, would seriously aggravate the situation.

Clearly, we should attempt to acquire a better understanding of the climatic changes of the recent past and their impact on our environment. We should also learn a great deal more about the causes of climatic change and attempt to assess in greater depth the relation between society, economics, and climate. We should then at the very least have some idea of what might lie ahead. At best we might be able to predict, even if imperfectly, the climate of the next 50 to 100 years and thus plan for a world that may undergo drastic climatic changes—perhaps within our own lifetime or that of our children.

Climate and the Oceans

Why discuss climate in a book about the oceans? In a general way the answer is obvious: weather and sea seem closely related, particularly on west coasts where the most important weather, the rain and the wind, come in mainly from the sea. Also, sailors and oceanographers, exposed so totally to the weather and so often frustrated and endangered by it, regard the air and the sea essentially as one. Coastal dweller and sailor experience the sea differently—the beach walker enjoys a stormy day, whereas the oceanographer prefers a windless one, sunny if possible but rainy if it must be—but for both, sea and weather are intricately related.

There are more solid grounds for discussing climate and ocean in a single context. In previous chapters we have discussed some of the interactions between the two. The role of the Gulf Stream, for example, in making northwestern Europe more habitable than northeastern Labrador, which lies at the same latitude, is well known. The oceanic circulation is a major force in transferring heat from low latitudes to the circumpolar regions, and its function as a heat carrier complements that of the air. The atmosphere has considerable kinetic energy, but it has a small mass and cannot carry much heat. It can, however, transfer its small heat content quickly from one place to another and can respond very rapidly to changing conditions. The ocean, on the other hand, moves very slowly but contains a great deal of heat; the uppermost ten feet of water alone contain the same amount of heat as the entire atmosphere.

Consequently the sea is a great stabilizer of temperature and carries vast amounts of heat over long distances from warm to cold areas. Moreover, once set upon a new course, the sea's inertia will maintain that trend much longer than the atmospheric circulation, which can shift rapidly back and forth. In some areas weather patterns appear to be related to oceanic conditions detectable months in advance; in 1962–63, catastrophic winter rains in California were foreshadowed in the previous summer by an eastward flow of large patches of unusually warm surface water northeast of Hawaii. Such relations between temperature anomalies in the central North Pacific and the weather on the Pacific coast six months later were first observed during the 1960s and lead us to believe that long-range weather forecasts may eventually be made based on oceanographic observations.

It seems likely that the Pacific coastal weather is influenced by changes in the Pacific oceanic circulation which, in turn, may have been induced by unusual weather conditions somewhere else, for instance in the equatorial zone. Since this close relation between oceanic circulation and climate is likely to be general, we can attempt to deduce past climates—which tend to have vanished leaving few traces—from past ocean circulation patterns recorded in oceanic sediments.

Extensive studies of ocean sediment cores dating from the last glacial period have recently provided a clear picture of the temperature distribution and surface water circulation in the North Atlantic and North Pacific during the glacial peak about 18,000 years ago. In responding to the subsequent climate change the two oceans showed very different behavior.

Third Beach, Newport, R.I. by Worthington Whittredge.

The waters of the North Atlantic today are quite warm, up to 50 degrees Fahrenheit, as far north as a line from Newfoundland to Iceland. Southward the temperature increases rapidly, and between Cape Hatteras, North Carolina, and Spain the water is approximately 70 degrees. During the last glacial period, the balmy Atlantic was quite a different ocean. The entire northern and northeastern Atlantic was an arctic sea with a permanent ice pack extending as far south as a line from the Carolinas to Spain and North Africa. South of the ice margin, the temperature increased very rapidly, and in the warm equatorial Atlantic conditions were quite similar to today (see Figures 6 and 6a in Chapter Three).

In the Pacific, on the other hand, the change from 18,000 years ago to present conditions has not been very dramatic. During the glacial time, the colder waters of the northernmost Pacific moved about 500 miles farther south—a small distance compared to the southward shift of ice in the Atlantic of 2,500 miles. Sea ice is now rare in the North Pacific and was rare 18,000 years ago except near the Alaskan coast and in the extreme northwest. Understandably, the coastal climates of the North Atlantic were a great deal more severe during the last glacial, whereas those of the Pacific coasts received somewhat more precipitation, but otherwise did not change drastically. It is not clear from this comparison whether the ocean caused the climate to change or the climate modified the ocean—a question that deserves further study—but it is obvious that the study of the old ocean can contribute to our understanding of past climates.

The sediments of the ocean floor preserve an excellent and often quite complete record of changes in the ocean. On land, by contrast, the fossil record of climate is spotty and incomplete. Floods, storms, and glaciers deform the landscape, but the traces of these events are generally fragmentary, difficult to correlate, and hard to assign to any precise time. The best information on ancient climates is contained in the fossil spores and pollen that have accumulated over the years in swamps and peat bogs. Painstakingly studied over many years, the pollen records chronicle a detailed story of the changing vegetation, which responds mainly to changes in climate: rainfall, temperature, length of winter and summer, and so forth. Trees also record droughts and temperature changes in the changes in thickness of the annual layers that build their trunks. Thickness variations of tree rings have been measured in trees of great age, including the California bristlecone pines, of which there are living specimens more than 3,000 years old. By matching overlapping segments of tree-ring records from trees of progressively greater age, the tree-ring chronology has been extended back almost 8,000 years. It contains a climatic record of great value, which at present is mainly restricted to the southwestern United States but is being expanded elsewhere.

Given the deficiencies of the climatic record on land, the archives of the seafloor assume great importance. They have been studied intensely in the last few decades and form the basis of our understanding of the climate before man began to measure rainfall and temperature with instruments a mere three centuries ago.

Climatic History of the Distant Past

Past climatic variations have occurred over a wide range of time scales —from hundreds of millions of years to centuries or decades. Major ice ages, each lasting millions of years, have appeared intermittently during the history of the earth. Although the only one that is known in detail is the most recent ice age, which started in earnest about 10 to 15 million years ago with a major glacial episode in Antarctica, the occurrence of another ice age, which glaciated large parts of Pangaea 300 million years ago, is well documented. This ice age lasted some 50 to 70 million years. Before that time, and within the last one billion years of earth history, there are indications of four more ice ages, spaced about 200 million years apart. Records extending back that far have been largely destroyed by the ravages of time, but there may have been another ice age more than two billion years ago. All except the most recent of these episodes (which may not be over yet) lasted 50 million years or more. Each ice age consisted of a series of glacial periods alternating with interglacials with a milder climate. Only for the most recent one do we know the sequence of glacials and interglacials in some detail (see Table 3, next page) but it is likely that the older ice ages had a similar fluctuating climate.

During both the present and the Pangaean ice ages, continents were located at or near the poles of the earth. Consequently many geologists have thought that a land mass in a polar position, or at least an enclosed ocean such as the present Arctic, is required to produce an ice age. Some have gone beyond this and have postulated that it is the arrangement of land and sea on the globe which actually causes ice ages and that ice ages are thus a product of continental drift. Given that we know the distribution of continents and oceans for only two out of seven (and perhaps more) ice ages, this inference is rather far-fetched and is perhaps an oversimplification. One of my favorite examination questions for graduate students in oceanography involves envisaging a scenario in which the South Pole has moved to a small island in the southern South Atlantic. This places the North Pole in the open Pacific southwest of Japan. If one then attempts to sketch the ocean circulation and climatic pattern for such a configuration, it turns out that the crucial question is whether or not one can have an ice cap on a pole in mid-ocean. The answer to that question is unobtainable with the present state of our knowledge. It clearly is premature to attribute

Table 3: The Chronology of the Latest Ice Age.

Time (in years)	Events
250-350 million	Pangaean Ice Age
50-250 million	Interval of warm and relatively uniform global climate.
About 50 million	Climate begins to deteriorate: poles become much colder.
10-15 million	First major glacial episode in Antarctica: Ice Age begins.
About 4 million	Arctic Ocean becomes ice covered.
2 million	First glacial episode on Northern Hemisphere begins.
1 million	First major interglacial. Two glacials, followed by interglacials.
110,000	Most recent glacial begins.
	Two glacial maxima separated by episodes of somewhat milder conditions.
18,000-20,000	Last glacial maximum.
10,000-15,000	Melting of ice caps.
10,000-??	Interglacial.
Future	Another glacial?

the cause of ice ages to continental drift when our understanding of the relation between ice caps, climate zones, and ocean circulation, on the one hand, and the position of continents, on the other, is so deficient.

The alternation of glacials and interglacials in the most recent ice age has sometimes been explained by changes in the orbit of the earth around the sun. This astronomical explanation, first proposed by the Yugoslavian meteorologist M. Milankovich 50 years ago and quite widely accepted ever since, has produced an interesting variation. Two Canadian earth scientists, J. Steiner and E. Grillmair, recently attributed the regular alternation of glacial and global warm periods to changes in the orbit of the entire solar system around the center of the galaxy. They computed a curve representing the changes in centripetal force on the sun that would follow from variations in distance to the galactic center and showed that ice ages always seem to occur during times when the centripetal force is minimal, presumably because of changes in the intensity of solar activity. The idea is interesting, but it rests on a number of explicit and implicit assumptions, including some rather courageous decisions on the time of occurrences of the individual ice ages and equally tenuous estimates of the past solar orbits around the galactic center.

Whether the ultimate cause of the ice ages is some galactic constant or continental drift will remain the subject of lively debate for years. In the interim we will have to be content with the somewhat better documented observation that the shifting of continents and oceans and the resulting changes in temperature and circulation of oceans and atmosphere exerted considerable influence on climate, even if they are not the primary cause.

The Last Glacial Episode

The story of the slow decline from the warm and globally even climate 100 million years ago to the beginning of the Ice Age 10 to 15 million years ago has been described in Chapter Two and summarized in Tables 1 and 3. The history of the early glaciations of this ice age is still obscure; however, we do have quite clear ideas about the last million years, during which there have been seven abrupt transitions between glacial and interglacial periods. Each of the glacial periods represented a time of intense cold, with ice covering large parts of the northern continents and most high mountain ranges of the world, and lasted about 100,000 years. Each interglacial period was much shorter—about 10,000 years on the average—and had a milder climate, with ice caps only in the polar regions and the highest mountains, and a smaller temperature range from the equator to the poles. The last of the glacial periods began about 110,000 years ago, during which there were three major advances of the ice caps, the last some 20,000 years ago (see Figure 8). When the ice cap melted between 15,000 and 10,000 years ago, the current interglacial began.

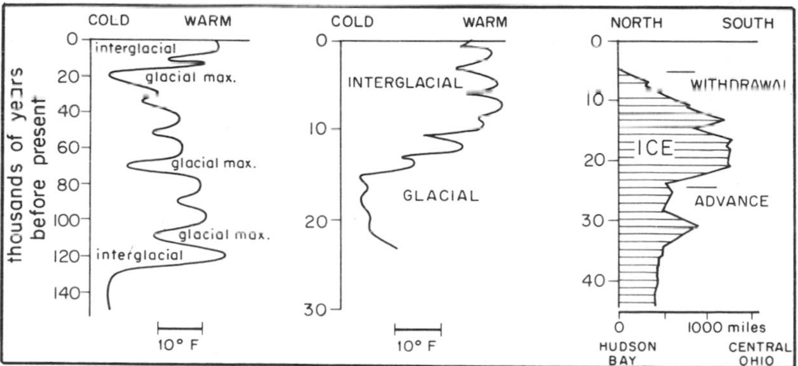

Figure 8: Variations of cold and warm weather in the Northern Hemisphere on various time scales. The alternation between glacials and interglacials is shown on the left: periods of very cold and of temperate weather alternate on a scale of tens of thousands of years. In the center we see the change from the last glacial to the present; on this much finer time scale we see variations between warm and cold superimposed on the transition but the differences between harsh and mild conditions are smaller and the time scale is shorter. The rapid advance and withdrawal of ice that occurred over Canada and the central United States is illustrated on the right. The data are from *Understanding Climatic Change,* a report by the National Academy of Sciences (1975, pp. 130, 154).

The last 25,000 years are of special interest to us. They include a major southward advance of the ice caps in Europe and North America, a rapid warming from 10,000 to 15,000 years ago, and a milder and relatively even climate since that time. This entire period is of interest, not only because it includes a good portion of man's march toward civilization, but also because the flora and fauna that existed then on land and in the sea were essentially like those of today, so that we can use their fossil remains to make interpretations of past climates and environments.

In North America the ice that covered eastern Canada and the eastern and central United States reached its maximum between 18,000 and 22,000 years ago, extending as far south as New York and central Ohio. The mountain ranges in the western parts of North America had a less extensive ice cover, which reached its maximum later—about 14,000 years ago. Although this last glacial episode was not the biggest, the extent of the ice cap in the Northern Hemisphere-including North America, northern Europe, the Alps, and Siberia—was probably about 90 percent of the maximum ever reached.

About 14,000 years ago, climatic conditions improved abruptly and widespread melting began. The retreat of the ice was very rapid; in the northeastern United States, for instance, the ice withdrew 1,500 miles in less than 5,000 years, and at times the ice receded several miles in a single year. About 10,000 years ago the ice sheet in the western United States had disappeared completely and present conditions, as far as ice cover is concerned, were established everywhere about 7,000 years ago. Only relatively minor changes in ice cover have occurred since that time, although glaciers have waxed and waned enough to have a major effect on the environment—and the populations—of their own valleys. The withdrawal of the ice was not uniform, but was accompanied by sharp and sometimes dramatic reversals which temporarily brought arctic conditions back to regions that had already been temperate in climate and vegetation for some time—such as the temporary return to tundra conditions in northwestern Europe between 10,000 and 11,000 years ago, mentioned early in this chapter.

During the period 5,000 to 7,000 years ago the climate was noticeably warmer than it is today. This period represents the climatic optimum of the post-glacial period, and evidence of prehistoric farms have been found in Scotland and northern England in elevations where today no agriculture is feasible. This climatic optimum terminated with a rather sharp turn to much cooler weather 5,000 years ago. One of my former professors, geologist C.H. Edelman, a man with a very wide-ranging mind who made important contributions in fields ranging from sedimentation to agricultural economics, was very much impressed with the climatic optimum and its sudden end. During a fascinating discussion one evening many years ago he suggested, quite seriously, that we ought to look more closely at this time, particularly in that cradle of human civilization, the valley of the Tigris and Euphrates rivers in Mesopotamia. As he put it, the fine conditions of the climatic optimum should have made a land of milk and honey of the upper valley of the Fertile Crescent and the surrounding slopes of the southern Caucasus—a veritable paradise where human existence was easy and survival a matter of no great skill or technology. "Imagine," he continued, "what must have happened when the climate suddenly, in a few decades or centuries, changed to the hot summers, droughts, and flash floods of today. It is easy to see the Angel with the Flaming Sword driving man from Paradise."

More recent archeological findings now place the origins of agriculture and civilization in the Middle East several thousand years before the end of the climatic optimum and therefore do not support Edelman's view, but it is an interesting way of looking at man's relation to climate. The story illustrates the tendency, not uncommon among geologists, to attribute to climatic variations (and all other sorts of geologic events) a major influence on man's affairs. Among historians, on the other hand, climate is not popular as a driving force of history, nor are earthquakes, volcanoes, and other geological forces.

Climate and History

In the study of climatic variations of the last ice age a variety of geologic records can be used—oceanic sediments, the deposits of waxing and waning glaciers, changing vegetation patterns recorded in peat bogs by pollen, tree-ring analysis, and cores of ancient Antarctic or Greenland ice. Most of these geologic records are difficult to quantify, and because of difficulties in obtaining precise ages, they do not allow us to separate events that closely follow each other; the best we can hope for is to ascertain an average climatic change over moderately long time periods of centuries or longer. This is obviously inadequate if we wish to relate climatic changes with historical events over the last few thousand years, although some of the best of our tools—in particular pollen and tree-ring records—can sometimes be dated with precision to decades and even years. Modern weather records, including temperatures, wind, and precipitation, are available only for the last century, but in a few places, such as England, some measurements date back to the late 17th century.

Fortunately, the historical climate record can be extended back over the last thousand years by a variety of ingenious means, although the results are fragmentary and geographically restricted. Hubert H. Lamb, a British meteorologist, has unearthed vast amounts of weather data from diaries, historical texts, chronicles, and, for the past three or four centuries, the bridge logs of British naval officers. Daily records of the weather were kept on British warships wherever the British fleet went—and British men-of-war covered a large part of the world even in the 17th century—and they have been dutifully filed at the British Admiralty. With this sort of data Lamb has been able to construct complete weather maps on a month-by-month basis for the eastern North Atlantic and the seas of northwestern Europe during the period from about 1700 to 1850 A.D.

In France, historian E. LeRoy Ladurie has scoured the archives of villages and churches for the records of the dates and quality of the grape harvests, which are very sensitive to the rainfall and warmth of spring and summer. Similar studies have been carried out in other countries that have a long written history—for example, the agricultural successes and failures in Sweden over the last 500 years and the climate and agriculture of Iceland for an even longer time. Such studies, taken together, enable us to deduce a fairly accurate climatic history for the Atlantic-European region since 1000 A.D. Because the material is fragmentary and qualitative rather than quantitative, there is considerable room for interpretation, and no two histories are exactly the same, as is shown by the climatic records for Paris and London compared with those for Iceland (see Figure 9). The general climatic trend, however, stands out quite clearly. After a somewhat inclement period around 1000 A.D., the climate improved rather significantly, and the period from 1100 to 1400 was quite warm, interrupted only by a few severe winters in Iceland, Scandinavia, and perhaps England early in the 14th century. At this same time France also had several wet winters and cool, damp summers, which resulted in devastating harvest failures and famines.

A prolonged climate deterioration began around 1450 that lasted until the first half of the 19th century. This period, known as the "Little Ice Age," brought long, harsh winters and short, wet, and cool summers to northern and western Europe and has been blamed for all sorts of social and economic misfortunes. It is widely assumed that the warm period preceding the Little Ice Age made it possible for the Vikings to settle Greenland and survive on an agricultural economy. The Little Ice Age, however, brought about a large southward advance of the ice in the Atlantic that extended well south of Iceland, interrupting communications by ship for many years at a time and destroying the agricultural base in Greenland. The resulting isolation eventually obliterated the first European colony in North America. Iceland, home base of the Greenland colony, experienced its own troubles during this time. Its wheat-based agriculture of the early Middle Ages came to an abrupt halt and had to be replaced with sheep and cattle grazing on the natural pastures of the virtually treeless island, but

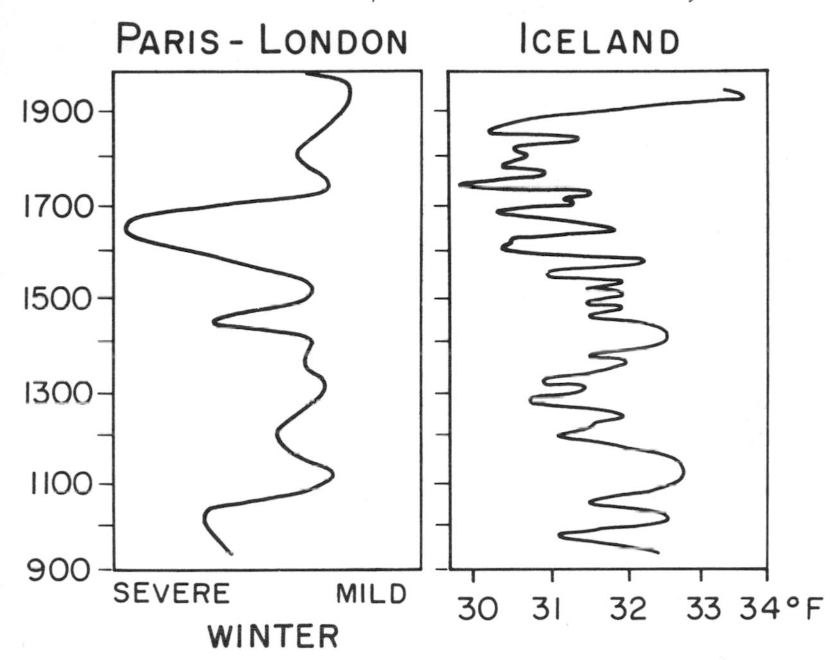

Figure 9: Climate variations in the Northern Hemisphere during the last thousand years. On the left is the variation in severity of the winters in Paris and London (from data in *Understanding Climatic Change*, a report by the National Academy of Sciences, 1975, p. 152), and on the right the variation of the mean annual temperature in Iceland modified from a review by Reid Bryson in *Science* (vol. 184, 1974, p. 755).

even this much less demanding economy was seriously and sometimes disastrously hampered by cold, wet summers and a marked decrease in the length of the growing season.

Frequent harvest failures in the 17th century may have been in part responsible for the military adventures of the Swedish king into Germany and central Russia to alleviate the unrest at home resulting from famines. Harvest failures and famine rather than the cussedness of the Scots may have caused the Scottish Border Wars. Also in the 17th century the glaciers in the Alps of France, Switzerland, and Austria began a major advance that destroyed villages and local alpine economies. In western North America, a similar glacial advance occurred which lasted well into the 19th century.

In the middle of the 19th century, the climate over the Northern Hemisphere began to improve markedly, and from that time until the 1940s the climate was extraordinarily favorable, stable, and without parallel in the preceding thousand years—perhaps even since the climatic optimum more than 5,000 years ago. However, climatic improvement in one region can simultaneously bring disaster in another, as is illustrated by what happened in the central and western United States. Here the very favorable weather with adequate rain that existed during the waning years of the Little Ice Age was replaced in the last three decades of the 19th century by extensive droughts, which very nearly ruined the recently established settlements formed as a result of the westward migrations of the 1850s. The disastrous droughts caused much Congressional concern over water rights, played a large role in the establishing of the United States Geological Survey, and greatly influenced the nation's assessment of the potential of the western territories.

That we have just experienced a period of, in the main, very favorable and, above all, dependable global climates is beyond question. This period, which peaked between 1920 and 1940, brought the Northern Hemisphere warm summers and mild winters unequaled since the climatic optimum and yielded weather patterns that were highly predictable. In the 1940s, however, a gradual drop in the mean temperature began, and since the 1960s the weather has become strikingly less stable—with a seemingly endless series of anomalous years nearly everywhere on earth. One need only remember the extensive droughts in central Africa and the absence of the monsoon rains in India during the early 1970s, the very hot summer of 1974 and delayed winter of 1975 in northwestern Europe, the exceptional drought in England and France in 1976, the anomalous wet winters of 1962-63 and 1968-69 on the Pacific Coast, or the droughts in California and the Middle West during the mid '70s to realize that these recordbreaking weather patterns indicate a profound change in conditions from the norm of the past century.

Geologist George Kukla of Columbia University has shown, based on satellite photographs, that in a mere three years, between 1971 and 1974, the total area of the world covered even in the summer with snow and ice has increased by more than 20 percent. Since ice and snow strongly reflect the solar heat back to space, this increase means an equivalent amount of uselessly reflected solar radiation. According to various estimates, the cooling of the Northern Hemisphere since 1958 has amounted to something on the order of one degree Fahrenheit or perhaps even more. Since the temperature increase from 1880 to 1940 was about two degrees Fahrenheit, we are halfway back to the Little Ice Age of my grandfather's time, when you could skate every winter. It is doubtful that the same cooling trend is occurring in the Southern Hemisphere; some observations that indicate a warming trend there have created considerable confusion in the debate about the causes and the likely future of the climate change we are now experiencing.

Temperature differences of a few degrees may seem rather small, and we shall examine more carefully below what constitutes an important climatic change. What is obvious, however, is the remarkable and rather dramatic change toward much less predictable weather with large variations from year to year between droughts and floods, cold summers and hot winters. Unfortunately, the historical records suggest that the present instability—rather than the predictable climate of the first half of our century—is the norm over the long term.

What Is a Significant Climate Change?

Much of this book has dealt with matters that are interesting to know for their own sake, and not much has been said of the practical application of this knowledge. When we talk about climate, however, the subject touches very directly on both our personal experience and well-being and on the welfare of nations. The most obvious application of such knowledge is climate prediction. The climatic variations of the past thousand years are likely to continue in the future, at least for the next century or more. Are variations of that magnitude important enough to justify the effort to predict them?

The answer is a firm yes, and it has been given by many students of climate prediction. Reid Bryson, a climatologist at the University of Wisconsin, is one of the advocates of intensified study of past and present climates, and I will draw some illustrations from a summary of his views in *Science* (vol. 184, 1974, p. 753).

Historical records of the presence or absence of sea ice on the coasts of Iceland can be transformed into annual temperature variations for the last 1,000 years (see Figure 9). Using these records as well as historical data on harvests, P. Bergthorsson has shown that between 975 and 1500 A.D., a period of generally favorable climate, there were 12 years of famine in Iceland, five of which fell in the period from 1250 to 1390. Between 1500 and 1800, during the Little Ice Age, there were no less than 37 years of famine, 34 of them in the very cold period between 1600 and 1800. Agriculture in Iceland depends on grass that feeds livestock; not since 1500 has even a modest wheat-growing effort been possible. Although the mean annual temperature differences between cold and warm times were small, ranging from two to four degrees Fahrenheit, their impact on the grass production was large. A drop of two degrees reduces the number of growing days per year by two weeks, a drop of four degrees reduces the growing season by 40 days, or 25 percent. If we also take into account the average temperature during the growing days, we find the growing season reduced by 27 percent for a two-degree drop in temperature and by 50 percent for a four-degree drop. In a subsistence economy like Iceland's, such small temperature variations may thus have a very large impact.

India offers another striking example of the relation between climatic change and prosperity. It has a monsoon climate with alternating rainy and dry seasons. When the monsoon rains stay away, harvests fail and famine results. Before 1920 the rainfall dropped to less than half of its normal value on the average of once every eight or nine years. Between 1920 and 1940, the mean interval between years of severe drought increased to 14 years, and during the period of most favorable climate, the 1930s, there were no years of disastrously low rainfall. Now the number of dry years is increasing, and the average frequency is approaching its pre-1920 value. Coupled with the large increases in population over the last half century, a return to the frequent droughts of the late 19th and early 20th century could spell disaster.

Currently there is no permanent ice cover in the Canadian Arctic because the winter snow accumulation is just slightly smaller than the summer melt. Cloudy, slightly cooler summers would decrease the melting of snow, and even slightly higher snowfall in winter would increase the accumulation—either would allow some of the snow to remain over the summer into the next winter season. Thus relatively small changes in winter or summer conditions could change this non-glacial region into one where an ice cap could begin to grow—just as a small change in temperature in the 17th century presumably caused the glaciers in the Alps to advance and destroy many hitherto viable mountain communities. Colder, longer winters or cooler, overcast, and shorter summers would also sharply decrease the number of growing days in the wheatlands of central Canada and the Great Plains of the U.S., and would eliminate our ability to export food to the rest of the world. If similar climatic changes occurred in the Soviet Union—and it is likely that they would—the stage would be set for major geopolitical problems. Large and unresolvable food shortages in countries such as the Soviet Union or China might well have consequences that would jeopardize the economic and social structure of much of the rest of the world.

Other effects of a deteriorating climate might be more restricted, but could still be highly disruptive. A few years of subnormal rainfall in California, similar to the droughts of 1975 and '76, do great harm to the agriculture of the state and result in numerous forest fires. Droughts lasting more than a few years would have more serious consequences. California has already tapped its groundwater resources beyond the sustainable limit, and importation of river water from the northwest would be politically difficult, and technically impossible on a short time scale. Thus a sustained drought would imperil the state's ability to support the industry and associated urbanization that now make it a key component in the U.S. economy. Unfortunately, tree-ring data for the past several thousand years

Cradling Wheat by Thomas Hart Benton.

indicate that periods of subnormal rainfall lasting from three to ten years are not without precedent. In fact, the precipitation of the last 50 to 60 years appears to have been anomalously stable and perhaps above average.

Similarly, the drought of 1976 in Minnesota and the plains states has caused great economic stresses that may be repeated in future years, and in early 1977 the central and eastern U.S. experienced an energy shortage and associated economic problems caused by the most severe winter in over a hundred years. Several such abnormally cold winters in succession would seriously deplete our already marginal energy resources. Obviously climate prediction over the short range of decades would serve an important function in allocating water and energy resources and in determining the capacity of any given region to sustain industrial and agricultural growth. Given the evidence for a destabilization of the climate, research in future climates would appear to deserve equal priority with long-range (six-month) weather forecasting.

The demand for weather modification is heard frequently these days. When weather modification involves cloud seeding and similar localized activities, the economic and political consequences are limited. But other schemes may have more far-reaching effects. A dam across the Bering Strait has been proposed that would prevent cold arctic water from entering the Pacific. The principal benefit claimed for this project would be an unspecified amelioration of the climate along the Pacific coast of North America, presumably by stabilizing and extending the zone of winter rains. Unfortunately, the consequences of so drastic a change in ocean circulation are completely unknown; the climatic improvement might never materialize, and major undesirable side effects cannot be ruled out.

In order to irrigate more southerly lands, the Soviet Union is contemplating complete southward diversion of all major Siberian rivers that presently empty their fresh water into the Arctic Ocean. It can be reasonably assumed that the resulting increase in the salinity of the Arctic Ocean would reduce the ice cover, perhaps in a major way. The reduction of the ice cover, in turn, would change the amount of solar radiation reflected back by the polar area and might also, in unpredictable ways, affect the circulation of the Arctic Ocean. The combination of all of these changes could have a major effect on climate that cannot be foreseen. Clearly, a much better understanding of the mechanisms of climate is needed before we can afford to engage in such grandiose schemes. At the very least, we ought to study the results of similar experiments carried out by nature itself that came about as the sea level rose and fell during the Ice Age.

The whole issue of climatic change has another side. Nature varies with time, and large changes in environmental conditions and in the flora and fauna come about by purely natural causes. Given our current concern with man's ability to destroy his environment, we are inclined to blame society and technology for all visible changes that seem detrimental to us. Some of these alterations in our environment, however, might well have arrived anyway—without human assistance. Reid Bryson points out that a moist period in the high plains and Rocky Mountains between 1850 and 1870 was followed by decades of drought, which greatly reduced the production of grass. The disastrous effect on cattle ranching in the second half of the 19th century is well known, but the drought would surely have had a similar effect on the ability of the region to support the original herds of buffalo. If one assumes that the region's feeding capacity for buffalo was approximately proportional to that for feeding cattle, the drought would have diminished the original buffalo herds by 50 to 75 percent without any overhunting by man. We tend to blame ourselves for more than we need to at times, and if we had been around when the dinosaurs became extinct, technology would probably have had to accept the responsibility.

To take the stand that man does not endanger the environment would be ludicrous. But, given that we intend to remain on this planet, it is equally ludicrous to desist from actions important for the survival and well-being of man, without establishing firmly that the observed damage is *our* fault and not the result of some *natural* change in the environment. Increases in the world population and in our individual demands on resources stress the environment to the limits of its tolerance and sometimes beyond. In order to maintain the balance it is necessary to determine precisely what the environmental limits are—and that includes the prediction of how they will shift in the future.

Present and Future Climates: Controversy and Confusion

Having been suitably impressed with the magnitude of climate changes in the past and with the possible impact of even small climatic fluctuations on the lives of ourselves and our offspring, we should now ask what causes these changes. That question is not easily answered. First, our understanding of the global processes that make the weather is still quite limited. Secondly, although we have some knowledge of how the climate varied in the past, this knowledge is purely descriptive and the dynamics of change are still mostly a mystery. Finally, and perhaps most importantly, just in the last century we have begun to modify the climate ourselves, inadvertently of course—and to an as-yet-unknown extent—but possibly in a major way. It is not too surprising then to find a great deal of controversy among experts with some claiming that a new glacial period is approaching and others discerning evidence for a warming trend and worrying about glaciers melting and the sea flooding the land. Some see the greater climatic variability of the last decade as a passing thing without much meaning whereas others interpret it as an indicator of major changes (usually unpleasant ones) that are around the corner. Some think that man is playing a new and perhaps dominant role in climatic change so that past experience has little value in predicting the future. Others concede man's contribution, but think it trivial and regard past and future changes as evidence of nature's playfulness. As one of my students once wrote in an examination paper: "God made the world according to the geology textbooks and saw that it was dull. So he added some variety."

What actually happens when the climate changes for better or for worse from a state regarded as "normal" by human beings with short memories? Can we tie together these confusing shifts—from rain to drought in one place, and from cool summers and mild winters to hot summers and dry winters in another—into some meaningful description of a general global trend? Not quite yet, but perhaps soon. In each hemisphere, the global weather is dominated by a single stream of strong westerly winds at mid-latitude at an altitude between 10,000 and 60,000 feet. On the equatorial side, this *jet stream* spins off high-pressure cells, whereas on the polar side low-pressure areas or depressions form that bring the winter rains to the American West Coast and to northwestern Europe.

The jet stream has waves—troughs and ridges perpendicular to the direction of flow—comparable to the standing waves in a fast-flowing stream. These waves are generated by mountain barriers like the Rocky Mountains, by underlying land or sea surfaces that are colder or warmer than the surrounding region, and by the dynamics of the flow itself. The ridges or wave crests of the jet stream permit the warm regime of the lower latitudes to penetrate farther north than elsewhere, whereas the troughs channel cold northern air southward. In mid-winter, for example, ridges lie over the warmer ocean areas where they produce a mild climate, and troughs occur over the eastern parts of cold continents and bring severe winters. The jet stream ridge in the eastern Pacific warms the West Coast from California up to Alaska, whereas a broad trough over eastern Europe brings arctic weather down to the plains of Russia and Poland.

The jet stream varies from year to year both in intensity and in position. When the velocity increases, the troughs and ridges become more widely spaced. At such a time, the Atlantic ridge moves eastward over western Europe and produces a warm winter. When the upper flow weakens, on the other hand, the spacing becomes shorter and the trough that normally lies over eastern Europe is displaced westward and extends as far as England; this configuration brings long and severe winters with abnormally high snowfall, such as the memorable winter of 1963 which disrupted transportation and industry in Great Britain for several months.

When the jet stream shifts southward, the flow weakens and the low-pressure areas bringing rain move south of their normal paths, leaving western Europe and northwestern America drier than usual and narrowing the equatorial zone of tropical rains. The opposite effects follow a northward shift and intensification of the jet stream. Because the jet stream is not concentric around the pole and can shift its eccentricity, the results of a change in latitude or intensity are felt in different ways in different parts of the world.

Thus our understanding of the relation between shifts in the position and intensity of the jet stream and seasonal weather patterns appears to be fairly satisfactory and has enabled meteorologists to prepare increasingly successful weather forecasts for several months ahead. Our understanding of the causes of the shifts in the upper atmospheric circulation, on the other hand, is dismal indeed. Moreover, because the temperature of the underlying land and sea surfaces plays an important role in positioning the ridges and troughs of the jet stream, long-term variations in the circulation and the surface temperature of the oceans are also important, and we do not understand these fully either. On land, changes in the area covered permanently by snow and ice, or human activities such as the clearing of large forests have an important effect not only on the radiation of solar heat back to space, but also on the temperature of the land surface.

The climatic record of the past million years suggests a certain periodicity in the alternation of cold and warm periods, to put a complex phenomenon in simple terms. Talking about periodicity is a dangerous thing; the human mind likes to see such regularities because they make events easier to remember. Unfortunately, periodicity of a complex phenomenon is easy to fake, especially when the timing of the events is somewhat uncertain so that we can shift them back and forth a little to make them fit better. Nevertheless there is some fairly solid evidence of periodicities of 21,000, 41,000 and 100,000 years in the global climatic record. They are perhaps not immediately obvious from the illustrations I have given here, but recently geologists James Hays, John Imbrie, and Nicholas Shackleton have rigorously analyzed cores of oceanic sediments and have shown that fluctuations occurred in the oceanographic conditions of the southern Indian Ocean over the past half million years with similar periods of 21,000, 41,000 and 100,000 years. These periods of recurrence are interesting because they are the same as those of variations in the incoming solar radiation that result from changes in the earth's orbit. It thus appears that the increases and decreases of solar radiation are responsible, at least in large part, for changes in ocean temperature and circulation and ultimately for fluctuations in climate.

There are also shorter climatic periodicities. The data for the last 10,000 years are detailed enough to show that cold weather conditions of the Little Ice Age type return about every 2,500 years. Their cause is, unfortunately, not yet known. A long-established periodicity is that of the intensity of sunspot activity, which varies on 11- and 22-year cycles. The sunspot variation may correspond to changes in the heat output of the sun, and a relation between sunspot and climatic cycles has been claimed but not proven. An intriguing observation is that there may have been a protracted minimum in the sunspot activity—perhaps even a total absence—between 1645 and 1715 A.D., and other minima from 1790 to 1825 and from 1875 to 1910. The coincidence of these minima with observed periods of severe winters in the Northern Hemisphere may well be more than accidental, but the issue of the effect of sunspot variations on the weather is far from resolved.

Climatic periodicities, if they are real and not some figment of our imagination, offer the best hope for predicting future climates. The job, however, is made difficult by the fact that the variations resulting from astronomical influences are not significantly larger than the normal year-by-year random variations of the weather. The problem of sorting out the various causative factors is further complicated by the fact that the ocean responds rather slowly to any change in external conditions and introduces a lag time of 20 years or so. Thus one needs very long series of precise measurements to establish the variation of climate with time and to separate the systematic changes from the random fluctuations.

Such long series of measurements we do not have. Nevertheless some brave souls have ventured guesses regarding the likelihood of climatic changes in the near future. John Imbrie has argued that since glacial episodes are separated by interglacials with an average duration of 10,000 years, we should be due for another glacial period in the next thousand years. The probability that this will actually happen during the next few generations, however, is less than 1 percent. On a shorter time scale a smaller fluctuation, such as another Little Ice Age, has a 10 to 30 percent probability of arriving in the next hundred years. Even if further research is not able to improve this forecast, it seems to me enough of a likelihood that we should discuss nationally what society ought to do in case such a moderately severe climate deterioration actually does occur. The necessary adjustment to our ways of living and working would take decades to implement and cost a great deal of money, and it is none too soon for serious consideration of possible courses of action.

Unfortunately, natural climatic variations may not constitute the whole story. Humans are now sufficiently powerful and numerous to have real impact on climate without even intentionally trying; here I am not referring to schemes to modify the weather, but to inadvertent activities such as filling the atmosphere with dust or carbon dioxide. Carbon dioxide comes primarily from burning fossil fuels such as petroleum and coal, whereas dust is produced by an enormous variety of human activities—plowing, deforestation, slash and field burning, construction, and industry. Each and every one of us works daily to make the atmosphere dustier.

These two pollutants have opposite effects on climate. Dust in the atmosphere reflects solar heat back to space and thus decreases the amount of heat we receive at the surface of the earth. Hence, if we fill the atmosphere with dust, the average global temperature should drop. This insight is not new: geologists have long speculated that natural dust in the atmosphere from ash blown to great altitudes by major volcanic explosions might lower the temperature of the atmosphere; they have even attributed the ice ages to it. This suspicion has recently received support from the observation that the last five million years, during the Ice Age, have been a time of unusually intense volcanic activity. Volcanic input, however, tends to be sporadic and most of the time there is little ash floating around in the air. Most ash comes from spectacular eruptions such as that of Krakatoa in the late 19th century, but great eruptions are relatively rare. The human dust input, on the other hand, is continuous and is steadily increasing; Reid Bryson has estimated that it is at least 100 to 150 times larger than the average volcanic input. Most of our dust production does not go very high in the atmosphere so that it rains out quickly, but what does remain suspended for a long time is probably about the same amount as the average load of volcanic ash. In combination, the volcanic and human dust may have been responsible for the temperature drop of about one degree Fahrenheit that we have experienced since the year 1940—provided that the dust volume estimates are not too far off the mark.

Carbon dioxide has just the opposite effect. It prevents the heat reflected back from the earth's surface from escaping into space and thus acts as a blanket or makes a greenhouse out of the atmosphere. Currently, the carbon dioxide content of the atmosphere is about 25 percent greater than it was a century ago at the beginning of the Industrial Revolution. The rate at which we supply it is increasing fast, however, and once we have burned all available petroleum and most of the coal the carbon dioxide content of the atmosphere will have increased fourfold compared to the natural state. Unfortunately we cannot compute exactly what effect this addition of carbon dioxide has on temperature, but an increase of about one-half degree Fahrenheit during the last hundred years is not unreasonable. Such an increase has not been observed (except perhaps in the Southern Hemisphere), and we must conclude that the effect of the addition of carbon dioxide to the atmosphere has been compensated by some other factor that has driven the temperature down. For the future it is not clear whether the

The Numbering at Bethlehem, detail, by Pieter Brueghel.

increasing carbon dioxide will cause the temperature to rise, whether the increasing dust will cause it to fall, or whether some natural fluctuation of climate will override both.

In trying to determine whether the great winters of the Little Ice Age will soon return, we are confronted with natural changes that may or may not be periodic. The extent and rate of these fluctuations are only partly known, but we can definitely document a cooling trend accompanied by much greater variability of the weather which set in after 1940 and is still continuing. This trend may be part of a natural cycle and, if so, could become predictable with further study of previous climatic changes. On the other hand, our rather uncertain estimates of the effect of human activity—even though they seem to conflict sharply at the present time—suggest that our influence is large enough to make a great deal of difference and perhaps to overcompensate for any natural trend that may exist.

Should we then study the meteorology of human dust and carbon dioxide inputs and forget about the ancient climates? It is not surprising that there is no consensus among the experts. Those that rely on the natural cycle predict an onset of a colder, less stable climate; those that believe firmly in man's self-destructive ability see either dangerously warmer or unpleasantly colder climates ahead. It is easy to catch the imagination by recalling the climates of the past, predicting a new Little Ice Age for instance. If carbon dioxide or dust increases are the causes of climate change, however, we cannot make such comfortable comparisons because we do not really understand what the dire effects would be, although dramatic events, such as a major sea level rise that would flood New York and other coastal cities, have been predicted.

Public statements of scientists and their public relations officers reflect the uncertainty of our predictions only too clearly. Read *Time, Business Week, National Geographic,* the *Los Angeles Times, Scientific American,* and you will encounter every one of these possibilities, usually presented firmly and convincingly—and the more dramatic the consequences the better. Scientists tend to be fascinated with their own data and conclusions, sometimes to the exclusion of those of their colleagues. And just like everyone else, they sometimes make statements on inadequate grounds; if they do not, reporters will generally omit their "wishy-washy" qualifications.

One thing is quite certain, however: The climate *will* change because it always has, and our grandchildren may well look at the Flemish winter landscapes of the Renaissance with more understanding than I used to. Is that the direction I believe the climate will take? Personally, yes, but a few more years or a decade of research should bring a more firmly grounded answer. In the meanwhile, we had better be prepared to add "climatic uncertainty" to all the other unpredictable risks of modern life.

Seascape by Thomas Moran.

CHAPTER FIVE

THE SALTY SEA GOES UP AND DOWN

One can say that, without salt, a supreme enjoyment of life is impossible. It is so necessary to man that he even symbolically calls that highest form of mental exercise, the humor, salty.

Pliny the Elder, 23-79 A.D., *Naturalis Historia* XXXI

THE WATER OF THE SEA is salty. This simple fact, like the blue of the sky and the green of the grass, forms part of the basic knowledge that even the very young acquire. That the sky is blue and the grass green most of us can see for ourselves as soon as we set our first hesitating steps outside. That seawater is salty, on the other hand, is a personal experience for only a few of the world's children, yet all become early aware of it. Why?

Salt is an essential component of the human diet. Without it, our life and that of most other animals is a biochemical and physiological impossibility. On land, salt is in short supply in nature except in deserts. And until recently it has not been readily available to man. Salt is thus one of those basic commodities that has made and destroyed empires, caused wars, pillage, and rape, and brought riches to the fortunate: not because a lot is needed—fewer than 100 million tons of salt are consumed in the world for all purposes, compared to more than a billion tons of cereals—but because that little is very essential. Initially all salt comes from the sea, whether it is now obtained from salt pans like those of San Francisco Bay or from salt mines. Before we learned to mine the rock salt deep in the ground, an accomplishment only about a thousand years old, the sea was by far the main and most visible source—hence the saltiness of the sea entered into our basic awareness long ago.

There are other simple yet very basic facts of nature. One of these is that the shoreline, where the salt is found, is perhaps nature's most definitive dividing line. Many natural phenomena are separated from each other by boundaries or, on a more abstract level, can be divided into categories, but the clearest boundary that nature draws is the silver line of surf and swash. Now you are at sea, now you are on land, and there can never be any question which is which. It is so clear and leaves so little doubt about its meaning that it inevitably conveys a strong sense of permanence. Of course we are well aware, some California coast dwellers painfully so, that coastal erosion makes houses fall into the sea and can destroy railroad tracks, highways, and entire city blocks during a single severe storm. Land can also be gained from the sea, sometimes rather rapidly. When Columbus discovered the Americas, most of the present Mississippi delta was still muddy offshore water, and the Dutch are widely credited, with only partial justification, with having wrested their entire country from the sea. These, however, are localized events and the outline of the land is—in our minds and on a human time scale—a very permanent thing.

The apparent permanence of the shoreline dividing land and sea is, however, far from real. We have already spoken of the large variations in the proportions of land and sea that have occurred over geologic time. During each ice age, the level of the sea was much lower than it is today because so much water was locked up in glaciers and ice caps. In former coastal plains, now submerged, bones of aurochs and mammoth have frequently been caught in fishermen's nets and oceanographers' dredges; early man was their likely companion, although we have not yet encountered his remains.

These spectacular events are not the only way in which nature changes the position of the shoreline. Smaller variations in sea level are almost continuous. They have occurred in the recent past and may well, in the view of many scholars, have had a significant influence on the fate of man and his civilizations.

The Changing Level of the Sea

The great sea level changes of the remote past dwarf those of the ice ages, for the ice ages never extended the Gulf of Mexico all the way into Canada or flooded one-third of the surface of Africa. But the ups and downs brought about by the freezes and thaws of the polar ice caps were rapid, and so close to our time that we can visualize their impact in terms of modern geography. They were witnessed by Stone Age man and must have impressed him greatly.

During the last glacial period enough water was converted into continental ice, mainly in North America and northern Europe, to lower the level of the sea approximately 400 feet. On some coasts, such as California, where the offshore bottom falls steeply away to the deep sea, the lowering of the sea added no more than a few miles to the existing land, but in other regions the coastal plains widened enormously. The glacial shoreline of the Gulf of Mexico was 100 miles south of the present one in the eastern Gulf and nearly 200 miles south of Texas. On the East Coast of the United States the locations of present New York City and Savannah were about 100 miles inland. In Europe the North Sea and the Baltic did not exist, and one could walk from Wales to Ireland. The island archipelago of Indonesia was incorporated into a land mass welded to Southeast Asia which was more than half the size of Australia. These temporary lands were mostly plains traversed by incised rivers and covered with open woods, grasslands, and steppes. Early man surely lived on the eastern U.S. plains, hunting the mammoth and collecting oysters whose discarded shells can still be found along the ancient shores. He also used the dry land of the Bering Sea to make his way from Asia to North America without the help of boats.

Between 15,000 and 20,000 years ago the sea reached its lowest point. It began to rise again almost immediately when the ice started to recede. The rapid withdrawal of the ice produced an equally rapid rise of the sea, and 5,000 years ago the sea level was barely 15 to 20 feet below today's. Fifty feet of rise per thousand years is not such a dramatic event, and the inhabitants of the coast of California quite probably were unaware of it. On a wide shelf such as the Gulf of Mexico, on the other hand, this increase translates into a very rapid landward movement of the shoreline. Along the Texas coast, during the time of fastest rise, the shore must have advanced inland at a speed of something like 200 feet per year. That kind of loss of hunting and fishing grounds must have been impressive and memorable even to primitive people, and I think it is quite likely that the experience was retained in the racial memory and in legends for thousands of years. The rate of flooding in the Persian Gulf between 7,000 and 10,000 years ago was equally spectacular, and before the Euphrates and Tigris rivers filled the valley with silt, the sea reached inland over hundreds of miles. Since the Middle East and Mesopotamia already had become settled by that time, it is not surprising that flood stories are so common in the heritage of the world's oldest civilization.

Around 7,000 years ago the rise of the sea slowed markedly because the largest of all ice sheets, the one in North America, had disappeared. Yet the rise did not stop completely, and since then the sea has risen another 60 feet. This continuing rise is somewhat of a puzzle because it does not seem that enough water was left in ice sheets elsewhere in the world to account for it. One can think of several explanations, but none is very satisfactory. When seawater warms up, it expands; and during the Ice Age it must have been somewhat colder than it is today. In fact, because of the addition of so much cold water, the ocean may have been coldest while the glaciers melted. Most of the effect, however, should have been in the surface layer only, and the best estimate of the expansion due to the warming that followed is woefully inadequate to account for the missing 60 feet. It is also possible that the Antarctic ice cap continued to melt for a much longer time than the other glaciers and even now may be losing volume. However, since this ice sheet is surrounded by ocean, its withdrawal cannot be measured by studying the area of surrounding terrain that was once ice-covered as we can do in Europe and North America. A withdrawal over a distance of 60 to 100 miles would be necessary to account for a sea level rise of 60 feet, and there is no evidence that this has occurred. The melting of Greenland ice does not provide an explanation because, being so much smaller, its ice cap would have to withdraw something like 500 miles to accomplish the job.

A great many people have been engaged in compiling the history of the rising sea level of the past 15,000 years. To do this, one obtains sediments, fossils, or peat from old shorelines and dates them, usually by means of a technique called radiocarbon dating introduced by Professor Willard Libby. Cosmic radiation produces, high in the atmosphere, a small amount of carbon (carbon 14) which is heavier than the usual kind. Carbon 14 is radioactive and decays in about 5,000 years. It enters the tissues of all living things as a constant and very small proportion of the total carbon. After the organism dies, the carbon 14 decays over time. By measuring the amount of carbon 14 still present in the organic matter (peat, wood, shell, or bone) the time elapsed since the organism died can be established quite precisely. Radiocarbon dating is very widely used in geology and archeology and provides excellent dates, precise to within a century or so, for materials with an age range of more than 15,000 years.

When we first started to use this technique to date old shorelines, the task seemed simple indeed. Unfortunately nature tends to be quite perverse, and in the last two decades many pitfalls have been discovered. One of the most serious problems is that the amount of cosmic radiation varies slightly with time, and therefore so does the amount of carbon 14 that is produced and enters the body. Furthermore the amount of carbon 14 found in older organic materials is so small that even very minor contamination with younger material—for instance a modern mold growing in ancient peat—can throw the date off enormously. Finally it has turned out to be surprisingly difficult to determine with confidence that a sediment, peat, or fossil really does mark an old shoreline or sea level. In fact, what is sea level? Is it the high tide mark, the position of mean low tide, some kind of average of both? Storms can wash shoreline sediments and shells onto higher ground; streams can flush materials out to sea. Thus we have not found as many good samples as we had hoped that are suitable to fix the position of some old sea level with a precision of better than five feet or so. I once stumbled, lucky as I sometimes am, on some old shorelines in Brazil, now located above the present one on a cliff coast. The old shores were marked by crusts built by a calcareous snail that happens to attach itself to rocks and prefers to be within a foot of mean low tide. On this coast, where the tidal range is small, the snails gave me some nice, firm dates for positions of the sea five to seven feet above the present level around the years 1600-1700 B.C., 800 B.C., 250 A.D., and 900 A.D. But most dates are not that precise.

Of course, we cannot automatically assume that it is only the level of the sea that goes up or down. Coasts can do this also; they rise and fall as a result of regional forces within the earth. Volcanic regions in particular have provided some striking examples, although generally the amount of uplift or subsidence is small and the rate very low. Moreover, sediments, such as the deposits from beaches, bays, and lagoons, from which we get most of our samples, are very loosely packed when they first are laid down. As they grow older, they settle and sink. Peat especially, a fine material for dating, can compress a great deal when sediments are piled on top of it. Thus, in order to use our findings to determine the change of the global sea level, we have to correct for the rising and sinking of the land and for the compaction of the sediments. Because the corrections themselves rest on assumptions and on measurements that contain errors, they are naturally not very precise and therefore reduce the quality of the data.

Still another process is involved in sea level changes. If we plot sea level change curves from a variety of places, we find quite significant differences (see Figure 10, next page). In fact, in some localities the sea appears to have gone down rather than up in the last 10,000 years. It took us a surprisingly long time to realize what all this meant. Obviously changes in the level of the sea itself must be worldwide, and local differences must therefore be attributed to changes in the height of the land. If we look carefully at Figure 10, we see that the areas where the sea appears to have withdrawn or the land to have risen are all in the north—in the region that was once covered with ice. On the other hand, the sea level has risen the most just outside the boundary of the former ice cap, for example in New York. The explanation is as follows.

During the last glacial period, sheets of ice many thousands of feet thick imposed a heavy load on the crust of the earth. Under this load the crust gave way, in part elastically like a rubber ball under pressure. When the load was removed as the ice melted, the crust bounced back in a rather

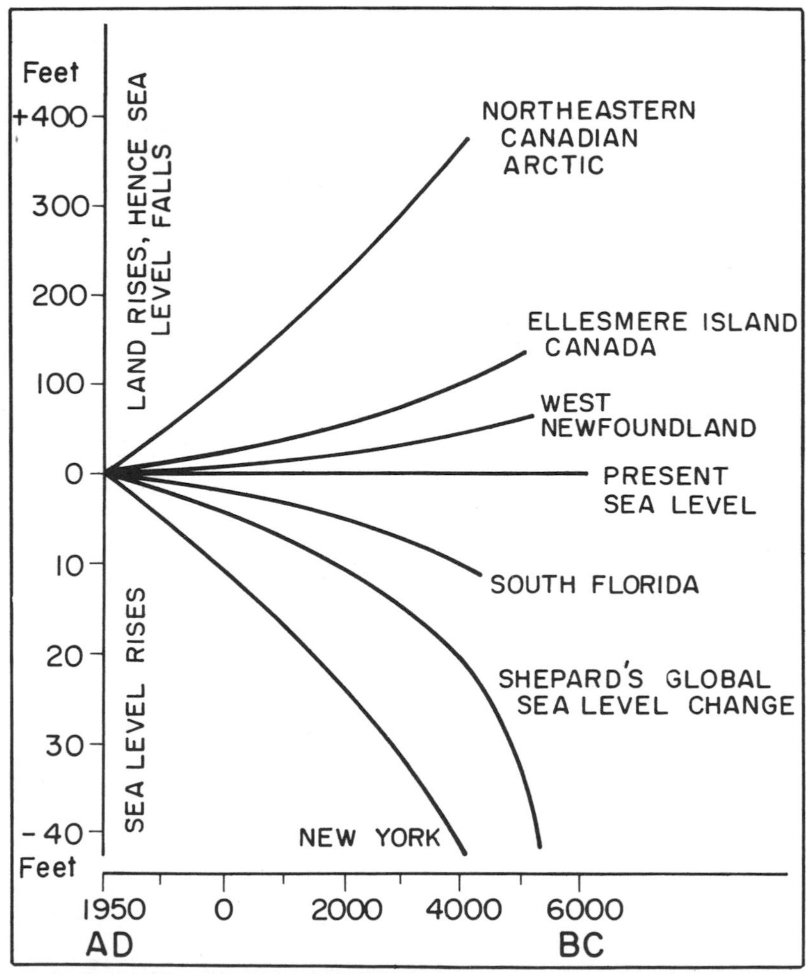

Figure 10: Changes of sea level in eastern North America during the past 6,000 years. Above the line indicating present sea level are coasts that rose because the land rebounded after the ice load melted away; the sea appears to have withdrawn. The closer the locality was to the center of the ice cap, the higher the rebound. Below the line of present sea level are coasts that sank a little but mainly were flooded because of the worldwide rise of the sea. Shepard's curve (constructed by Francis P. Shepard of the Scripps Institution of Oceanography) is an attempt to average various subsidence rates due to sea level rise and sinking of the land.

short time. That, however, was not all. The earth is not perfectly rigid, but acts under great pressure like a very viscous fluid, which flows so slowly that it is not noticed. Nevertheless the ice load was so large and had endured for so long that the earth responded with a flow of material from under the load to the areas beyond the ice margins. Thus the loaded region sank and the periphery rose.

After the ice melted, the reverse took place; the once ice-covered region rose several hundred feet and the periphery sank to restore equilibrium. The net effect was a rather large vertical change under the relatively small region that had been covered with ice and a much smaller one in the opposite direction over the much larger peripheral zone. Because the viscous flow deep in the earth is so slow, the process took quite a long time—on the order of many thousands of years—and has not yet been completed in the center of the once ice-covered regions, such as Sweden and Norway. Thus whereas eastern Canada rose for a long time after the ice had melted and still does so today, the coast from New York to Florida sank. Only by the time one gets to southernmost Florida are we outside the region where this compensation took place.

Naturally this vertical movement of the land complicates our deductions regarding the amount of worldwide sea level change since each curve sums the two effects. Thus a standard curve, which represents the global sea level change, makes little sense unless it is based only on data from outside the region where the vertical movements of the land took place. Unfortunately our present curve is not that impeccable, because many of our best dates come from places not too far from the ice margin.

It is still controversial whether, taken together, all of these effects explain adequately the 60 feet that the sea level appears to have risen in the last 7,000 years. As usual, the explanation will eventually turn out to be a combination of many factors. Yet, although this is a common experience in science, time and again we happily jump on the bandwagon of a simple explanation when a difficult problem turns up, because it is so much easier to live with a single cause. The history of science, and of geology in particular, is full of discarded simple ideas.

Sea Level During the Past 5,000 Years

For many geologists, the history of the earth ends and the present begins 5,000 years ago. At that time, 3000 B.C., the sea level was perhaps 10 feet below its present position, and the remaining rise has been too small to produce significant geological consequences or even, in general, observable ones. It is convenient to assume that the sea either reached its present level several thousand years ago or is still rising slightly. After all, these few feet are about equal to the precision we can get out of our samples. Thus, the cautious argue, it is wasted effort to try to squeeze more information out of our data.

Others, although recognizing the small geological importance of the issue, point out that to shore dwellers five feet is not insignificant. They have gathered information that suggests the sea rose episodically rather than smoothly and that, at times, it may even have been three to six feet higher than the present level. The first person to marshal evidence on this subject was Professor Rhodes Fairbridge of Columbia University. Fairbridge is a very colorful man with a wide-ranging curiosity and has expressed strong, often unusual views on many geological problems. Such people are often regarded with skepticism by those more narrowly specialized colleagues who believe that only experts should talk about a subject, and that one cannot be an expert in more than a few fields at once. Fairbridge's sea level curve, which is quite complicated with several highs in the last 5,000 years, was therefore not greeted with noticeable applause. Given my luck in finding several high sea level positions in Brazil, I tend to be on his side in principle, but some of the data that support his conclusions are not too sound.

Fairbridge's curve, however, has inspired a good deal of research in the last 15 years—the earmark of a useful scientific idea—and a significant number of carefully documented studies have been published. The pattern that emerges remains confusing. In some regions such as the North Sea, the Baltic, and the west coast of France, the observed sea level fluctuations are indeed episodic. Highs and lows, never more than three to six feet from the present mean sea level, appear to have occurred on the average every 200 to 500 years. Data from some remote corners of the world such as New Zealand and, oh yes, Brazil, support this. On the other hand, a very detailed study in eastern arctic Canada shows only a steady rise. On the East and Gulf coasts of the United States there is probably not enough information to be sure, but what there is gives no evidence for levels higher than those of today.

We have no good explanation for these small fluctuations of the sea level. Neither the rebound theory that we have discussed nor settling and compaction of sediments can account for them, because they would cause movement in one direction only instead of oscillations. Not all students of past sea levels believe that in the past 5,000 years the global level of the sea ever rose above its present height. Most of these nonbelievers consider the ups and downs as purely local phenomena, due to small vertical motions of the land or changes in the tidal range or the frequency of storms. Particularly in the Mediterranean, a region of common upheavals of the earth, it is easy to discredit any general sea level change curve by postulating local uplift or subsidence.

However, if the causes are local, why then would the same ups and downs occur at the same time in such remote places as New Zealand, Brazil, and the North Sea? Rarely do the known vertical movements of the earth extend over distances even as large as those from the eastern Baltic to the west coast of France. Of course, one can argue about the accuracy of the dates: Were the events in different places really synchronous? Why does the height of rise or depth of fall vary from place to place? Is it not easier to account for the severe floods in the Low Countries around the North Sea during the 15th century by means of exceptionally bad weather at the onset of the Little Ice Age? Do these events, which started the Dutch on their career as builders of dikes and reclaimers of land from the sea, actually have anything to do with a rise in sea level? If not, then what do we make of evidence that suggests that similar encroachments of the sea on the land took place elsewhere at the same time?

The geological record is too fuzzy to answer these questions. There is, however, a good deal of archeological information, often rather precisely dated. Medieval cities such as Haithabu in northern Germany and Reimerswael in the Netherlands are now permanently under the sea. In the Mediterranean many ancient harbors, temples, and whole cities are now submerged under several feet of water, whereas docks and sewer outlets of other ages are several feet above the present sea level. The archeological evidence in the Mediterranean ranges from several thousand years B.C. to well into the first millennium A.D. The list of gains and losses of land all over the world is long, and for each one we could perhaps find a local explanation.

However, if it could be demonstrated that shifts in the relation of man's feats of engineering to the level of the sea are synchronous over large regions, it would be difficult to find solace in local causes. Today the existing literature can provide just about any opinion that one wishes to find, and I regard the question as quite unsettled. Nevertheless, those who claim there were systematic sea level changes have an impressive array of data in their support. A summary of their views in Table 4 (next page) presents some of the more illustrious historical events that are often cited as confirmation. Hundreds could be added, but for each the opposition will gladly locate another ruin somewhere else that is either inappropriately above or uncomfortably below sea level in contradiction of the trend.

If we could think of a good cause for an oscillating sea level, we would be a good deal better off and would have something real to argue about instead of imprecise changes in levels of a few feet or insecure dates that can be shifted decades or even centuries either way. The latter is a particularly sore point because the claimed rises and falls are separated by only a few centuries.

Table 4: Changes in Sea Level during Historical Time and Events Caused by Flooding or Withdrawal of the Sea.

	Sea Level	Historical Events
A.D. 2000 —	normal	
—		
—		
—	higher	Extensive flooding of the land in low countries along the North Sea. The Dutch begin dike building.
—	lower	Low salt marshes exploited all over Europe.
1000 —		
—	higher higher	
—	lower	Ancient port of Ravenna landlocked and replaced by low-lying Venice.
—	higher	
—	normal	
0 —	(higher?)	New port constructed inland south of Haifa (Israel).
—	normal	
— —	lower	Many Greek and Phoenician ports constructed around this time are below sea level today.
—	(higher?)	
1000 —	higher	Ramses II builds first Suez canal; this would have been difficult to do with present sea level.
— —	lower	
—	higher	Coastal forests in Britain are inundated by the sea.
—	higher	
2000 —		
—	lower	
—		
—		
—		
B.C. 3000 —		

Professor Rudi Bloch of Ben Gurion University, who has, as we shall see, a stake in the subject of sea level rises, has come up with a particularly ingenious mechanism to explain small sea level oscillations. Bloch's early career was in salt chemistry and the salt industry which, in Israel, mainly uses evaporation ponds in the Jordan valley. He had noticed that when algal growth or dust darkened the water in the ponds, the increased absorption of the sun's radiation resulted in much more rapid evaporation. This principle is no novelty, but next he came across an unusual application at a hydroelectric station that drew its water from a melting glacier. During an unusually cool and overcast summer, the meltwater runoff was much below normal and threatened the power output. The problem was solved by dusting the glacier with a dark colored powder, thus increasing the absorption of heat. The Russians occasionally use this technique to hasten the spring thaw in frozen harbors and ship channels. With these facts in mind Bloch suggested that the melting of Antarctic ice—or Greenland ice for that matter—would have been speeded up a great deal every time an enormous volcanic eruption deposited an ash layer on the ice.

After a diligent search he located literature describing ash layers found in long cores of Antarctic ice that reach back thousands of years. Miraculously the ash layers did coincide in age with some of the high sea level stands that Bloch was particularly anxious to confirm. As far as I can tell, however, the idea has not been warmly received by others. Calculations by Hubert Lamb in England seem to suggest that even a volcanic eruption the size of that of Krakatoa would not be nearly enough to melt the water needed for a six-foot rise in sea level. Bloch disagrees, and because of inadequate knowledge of what precisely happens when you dust ice with ash, the problem cannot be solved.

There the case now rests, and it might be well to leave it alone, draw an average sea level curve, and forget about the oscillations as events that are, at this time, unknowable. That would be a pity, however, because by these oscillations of the sea hangs a fascinating tale. The tale derives, once again, from Bloch, whom I admire as a true example of a great scholar, and it weaves the sea level changes, the production and trading of salt, and the vagaries and vicissitudes of human civilization into a spellbinding story. It is good to be aware of the weakness of the underlying evidence, but the story may still well be true.

Salt, Man, and Society

Because no effort is required today to obtain the amount of salt necessary in our diet, we seldom realize how critical this component is. In reality, however, it is as essential as water. Without salt or with too little, life processes are impossible, and we can die from salt deficiency as easily and as unpleasantly as from thirst. An animal or human being suffering from a severe salt shortage is instinctively driven to satisfy his need. The drive is as strong as that of a drug addict and leads to equally aggressive, impulsive, and irrational forms of behavior. In today's world, with the exception of some primitive and remote populations, the acquisition of salt is no problem. Its wide availability and low price have obscured the once dominant role of this simple chemical. Traces of its impact on human behavior and history now remain only in traditional expressions, such as "a salty humor," "the salt of the earth," "sitting below the salt," and many others that testify to a role for salt that is now long gone. Economically, too, the small demand for salt and the large and cheap supply have removed it from the list of conspicuous world commodities. Extraordinary circumstances only rarely remind us of the key role of salt. During World War II, the island of Jersey off the Channel coast of France was occupied and heavily fortified by the Germans. The resulting isolation from the

Departure from Venice, detail.

mainland destabilized the salt supply and the fortifications cut off access of the civilian population to the sea. A black market in salt drove up the price of seawater to one German mark per hundred gallons.

Normally, of course, we cover our salt needs by going to the grocery store and spending a few nickels, giving the matter no thought at all. This was not always so, and a fundamental change from a major preoccupation with the salt supply to the present attitude of no concern came only as recently as a few centuries ago. The change was brought about by the ability to exploit rock salt in mines and the availability of the required technology and energy. As this industry grew and cheapened the salt supply, salt lost its preeminent economic role, and governmental and private salt monopolies, taxes on salt, and the salt trade—all still very much in evidence in the 18th century—gradually disappeared. It is curious today to note that part of the social unrest leading to the French Revolution stemmed from salt duties, the gabelle, which were regarded as unreasonable.

Before the 18th century, and much later in some parts of the world, salt was a major trade commodity. Suppliers of salt held dominant economic positions, and salt-poor nations or communities were underprivileged in the same real sense that countries with few raw materials or a low level of technology are destitute today. Salt was a basic form of currency; it was one of the key items in the African slave trade until less than a century ago. In the 18th century a few ounces of salt would buy one or two slaves in the west African slave market. Salt production and the salt trade required strong protection against raiders and quite commonly evolved into highly militarized and sometimes very adventurous enterprises. Most of this has vanished now but traces of the historical role of salt can still be found in remote parts of Africa or in the jungles of South America and New Guinea.

How did man acquire his salt before the advent of salt mines and modern salt ponds? The earliest and most basic supply came from food. Plants have a rather low salt content, but herbivorous animals consume such large quantities of vegetation that they are able to supply their own needs almost entirely from this source. In turn, meat and blood constitute the original salt source for predatory animals and for man. The salt supply of our remote ancestors, and that of some primitive tribes even today, came from eating meat and other, less appetizing dishes such as cheeses made from blood, milk, and cattle urine. It was also discovered early that salt could be manufactured by burning large amounts of vegetation and leaching the ashes. If the vegetation came from salt marshes, the yield was particularly good. Salt was also available, in impure form, in desert salt flats and in natural evaporating basins along the seashore.

At the seashore, there are two basic means of acquiring salt, both based on very ancient technology. On the coast, natural ponds with limited access for seawater can be used (or artificially modified for use) as evaporating ponds, provided the climate is dry and there is plenty of sunshine available. In low latitudes this was, and in many places still is, the main technique for salt production. In more northerly climates, the summer is too short and the rainfall often too heavy to permit the making of salt by evaporation. There a different technique of making salt by burning and leaching the peat of salt marshes was invented thousands of years ago and practiced until well after the Middle Ages. The Baltic and British salt trade was largely based on this source. Both types of salt making had the advantage that they took place in regions generally kind to man's survival, where hunting, fishing, and gathering, and later agriculture and trade, were possible and could support the salt industry; and transportation by water was cheap and comparatively easy.

The other main source of salt at the surface is in desert regions where there are naturally occurring evaporating ponds and salt flats. Although the salt here is easily acquired, albeit in impure form, the desert salt sources have the disadvantage that the region is not otherwise capable of supporting human settlements. Thus, trading is essential: a surplus must be produced to be exchanged for essentials such as food and sometimes even water. In addition, since desert salt basins usually exist precisely because they have interior drainage and are not connected by river with the rest of the world or the sea, all transportation must be over land and commonly over mountain ranges.

As the population increased over the millennia and the food supply inevitably changed from a dominant meat diet to a cereal-based one, the world divided into haves and have-nots as far as salt was concerned, the haves being either at the seashore or in interior desert valleys. In the economic struggle between the two groups, however, the exploiters of the desert source were seriously disadvantaged compared to the competition on the coasts because of the inhospitable desert environment and the dangerous, difficult, and costly trade routes. Thus whenever and wherever possible, the salt trade was focused on the coastal regions—and this is where sea level becomes important.

I might add here that salt has other uses besides that of a basic food additive. In the days before refrigeration and other modern technology, salt was absolutely essential for the preservation of meat and fish, olives, and even fruits and vegetables. During World War II my mother returned every summer to the ancient method of salting beans and other vegetables in large crocks for winter use, a rather unappetizing product that we thoroughly disliked. In pre-canning and pre-freezing days, food preservation for transportation or for nonproducing seasons rested almost solely on salt. Today salted fish is a specialty; one hundred years ago it was a necessity. Other uses of salt, such as the salt-curing of leather, no longer exist. All of these uses of salt were as essential for the survival of ancient societies and as important in their strategic, economic, and political life as petroleum is today. Moreover, substitutes for petroleum exist, but for salt there are none.

Salt Economy and Sea Level

It is therefore not surprising that until quite recently the manufacture and trading of salt were the principal reasons for the existence of entire cities, states, and sometimes civilizations. Others, controlling major transport routes, derived their income from the protection of the trade and from the levying of salt taxes. As a result, in the major salt centers of the ancient and medieval world—the Mediterranean coasts, the Jordan valley, and the Baltic and North Sea regions—economy, politics, and strategy were controlled largely by salt, with the seashore producers easily having the upper hand because of their favorable location.

It is the heart of Bloch's thesis about the salt trade and sea level that the coastal salt production is highly vulnerable to rapid sea level changes even if they are small or of short duration. It is fairly easy to see that the making of salt from vegetation or peat in coastal salt marshes located at sea level can be completely wiped out by a sea level rise of five or six feet. The ups and downs of the British salt industry as a result of the flooding of salt marshes in medieval times are well documented in chronicles and other historical records. Similarly the complex system of ponds and overflow channels that constitutes a seashore salt works capable of producing reasonably refined table salt from the mixture of useful and useless salts that seawater contains is precisely adjusted to a given sea level and is extremely costly and difficult to relocate when sea level changes. Thus during a rise of the sea level, especially a rapid one, the economic advantages of the coastal regions were sharply and suddenly reduced, and a near-monopoly in salt production and trade moved to the interior desert valleys and to the people who inhabited the surrounding mountains and were in a position to control the trade routes.

This relationship between the changing level of the sea and the shifting of power centers is particularly clear in the eastern Mediterranean where the people of the shore and those of the Jordan valley competed for the salt trade. From time immemorial, the passes from the Jordan valley to the coast have been controlled by cities such as Jericho, Jerusalem, Bethlehem, Samaria, and Hebron. Jericho is the oldest one of these with a known history as a city that extends back more than 8,000 years; its history shows a striking parallel with the changing sea levels shown in Table 4.

African Beach, detail, by Mariano Fortuny (1838-1874).

The Jericho region was densely populated and the city highly prosperous around 4500 B.C., but it became more or less deserted during the fourth millennium B.C. when the sea was low and the coastal cultures flourished. Then, between 2700 and 2200 B.C., the coastal zone declined and power moved back to the Jordan valley. Around 2000 B.C., when the level of the sea dropped once more, there was the great blossoming of the Aegean civilization of Crete and the presumed source of the Atlantis legend, Thera (Santorini). The Jordan valley became the realm of desert herders, and the cities of the mountain ranges barely held their own.

And so it went: a blooming of the inland cultures around 1700 B.C. when the sea was high; a renewal of the vigor of the sea people, this time the Myceneans, between 1500 and 1200 B.C. when sea level was low; then another focusing of civilization in the inland valleys and Mesopotamia as the sea rose once more. Finally came the great Greek civilization and later the power of Rome, both truly sea-based empires, both most powerful during long periods of stable, low sea level. The advent and expansion of the Greek, Macedonian, and Roman empires all were associated with a flourishing marine trade and salt industry at low sea levels of long stability, in counterpoint to the land-based world powers of Egypt, Persia, and Media, all of which possessed inland salt resources. The failure of the

Persian wars against Greece and the glory of Athens that followed can thus be seen as a defeat for the sea as well; it sank and was conquered.

Rudi Bloch has written a beautiful paper (unfortunately for many of my readers it is in German) in the journal *Saeculum* (vol. 21, 1970, p. 1) in which he views much of the history of the Mediterranean and a good deal of the medieval history of northwestern Europe in the light of sea level changes and their effect on the salt trade. He provides a great number of intriguing and often curious examples, and many of the aspects he reviews—such as the examples already cited or his discussion of the Baltic and North Sea trade of the Hanseatic League—are key historical events. His is unquestionably a one-sided view, and there are of course a great many factors besides the salt trade that influenced the rise and fall of the great civilizations of the past. It is often difficult to document the postulated relationships: we have seen how tenuous the sea level curve itself is, and the situation is little better with the complexities of archeology and history. Be that as it may, it is a refreshing view on a very old subject that I find highly thought-provoking.

Whether or not Bloch is entirely or partly correct matters less, it seems to me, than that he sets us to think and that he illustrates the height to which true scholarship can rise. From the moment I first met him I fell under the spell of this man who began as a salt chemist and traced his subject through geology and meteorology, archeology and history; who found his data in salt marshes and polar ice, in the Talmud and archeological excavation reports. Even the fact that he discovered that there might be, perhaps, such a thing as an oscillating sea level or that he turned to hypotheses about the periodic melting of Antarctic ice when his challengers demanded causes is unusual in this time of high specialization and low communication. I consider it both admirable and wonderful, in a world where the various scholarly disciplines are surrounded by signs saying "Only PhDs in this field may enter," to find a man who encountered an idea and tracked it wherever it took him, regardless of what area of knowledge he had to master.

A *Sea Flight* by an anonymous English artist.

CHAPTER SIX

PERILS OF THE TREASURE HUNT

On the bottom of the sea there is glitter of gold
Rubies and diamonds and treasures untold.

Folksong

THE SEA HAS BEEN USEFUL to man from the day he first set foot on its shores. From earliest times, fishing, hunting, and shipping have been part of his existence; they are still important for the economy of individuals and whole nations. The list of uses and resources of the sea is long and continues to grow. Offshore oil fields equipped with enormous installations for the production, storage, and transportation of petroleum already are a common sight in many places. The noise of man's activities in the sea, transmitted in the water over long distances, is beginning to interfere with the Navy's ability to listen to passing submarines. The Department of Commerce has estimated that by the year 2000 the U.S. may have more than a trillion dollars invested offshore, an amount that represents a significant part of the nation's total assets.

In the last few decades, as our awareness and knowledge of these three-quarters of the surface of the earth have rapidly increased, many novel prospects for what might be done with the ocean have occurred to scientists and promoters. John Isaacs of Scripps has suggested that fresh water in the form of icebergs could be floated from Antarctica to Los Angeles to supply drinking water. John Craven of the University of Hawaii has dreamed of floating cities, and others have talked of fertilizing the ocean or harvesting the power of waves, tides, or thermal layers.

Personally, I consider many of these proposals just dreams, and most of the rest a blend of vision and premature promotion. Although it is likely that the world will eventually see some entirely new uses of the sea, including floating cities and perhaps even submerged ones, the day is not as close as enthusiastic newspaper writers or aerospace companies looking for new and lucrative government contracts would like us to believe. Other prospects, however, may become a reality before I shall be too old to appreciate them; this seems to be especially true for the exploration of marine minerals, known in Washington parlance as "nonliving resources." Because petroleum is so much in the limelight already, I will limit my discussion of these marine resources to the hard minerals.

The thought that the sea might be good for something other than ships and fishing—beach recreation was not then as highly regarded as it is today—occurred to a distinguished French academician late in the 19th century. Based on a small number of not very precise chemical analyses, this gentleman estimated that there was enough gold dissolved in the sea to make its recovery an attractive proposition. A gold rush followed, on paper at least, which yielded a lot of inflated reports about the gold content of seawater in such places as the English Channel. An enterprising German even spent some frustrating years trying to devise a way to pay off the German debt from World War I with gold from seawater. It is quite amazing how people virtually anywhere will respond to the mere mention of gold. Not long ago, one of my colleagues mentioned in passing that, by his calculation, the seafloor muds in the eastern Pacific contained millions of tons of gold, and thereby provoked a positively greedy reaction from listening students and faculty. A few years ago, a company for whom I served as a consultant was looking at some possible gold placers offshore of the Alaskan coast, a harmless idea which came to nothing. Unfortunately word of their interest leaked out just as the company listed its stock on the New York Stock Exchange, where it immediately rose to dizzying heights—for a very short time. It took the management years to live this experience down.

The problem with gold in seawater is that it is so finely dispersed and so costly to concentrate that the effort is simply not worth the trouble. Thus the sea is a place where gold exists, but it is not a *resource* (i.e., a usable source) of gold—an essential distinction often forgotten in the rush for minerals. Seawater is indeed a resource of some other elements, and industry has found ways to exploit some of the major chemical constituents, just as salt has been exploited for millennia. Until quite recently a flourishing industry existed to extract bromine, an essential component of leaded gasoline, from seawater, but the availability of other bromine sources coupled with the reduced consumption of leaded gas has led to the demise of this business. Similarly the sea was for decades a major source of the light metal magnesium, but again other sources and a diminishing demand have eliminated the commercial viability of extracting magnesium from seawater.

The notion that mineral resources might exist on the ocean floor came much later, in the early 1960s, as a result of our rapidly increasing knowledge of the geology of the ocean. In less than two decades, it has transformed the profession of ore geology; where there was once almost total disregard of the ocean, there is now rapt attention. The whole course of events would be an excellent illustration of how pure and selfless research ultimately produces benefits to mankind ranging from refrigerators to heart valves, except that thus far virtually no money has been made in marine minerals. The story of our hunt for the sea's mineral treasures is thus rich in lessons, but whether large material rewards will ultimately give it a happy ending is still unknown.

A Plethora of Nonliving Resources

As I have sketched in the Introduction, in the early years of the 1960s we witnessed a sudden blooming of our nation's interest in the oceans, highlighted by the establishment of the National Marine Council. The excitement about the potential of the sea was composed in roughly equal parts of sound recognition of the sea's real importance, self-serving attempts of scientists to justify their demands for a larger budget, and bureaucratic jockeying for power in what was suddenly perceived as a major new horizon for government empire-building. Industry contributed vigorously to the dialogue, but mainly by offering (expensive) services, claiming (unsubstantiated) expertise, and clamoring for government contracts. The investment of private venture capital was relatively small and is so to this very day, with the exception of the petroleum industry and one or two large mining companies.

All the activity produced a remarkable range of assessments of the pot of gold at the end of the rainbow. The director of the Bureau of Mines presented some estimates to the National Marine Council based on what he called the "mirror image concept": take the area of the offshore shelf, delineate a comparable area of land onshore and determine its mineral value; then apply the mineral value so obtained to the offshore lands. If it is high, it is an acceptable estimate of offshore nonliving resources. Such estimates, however, totally disregard the fact that continental shelves are fundamentally different in composition and origin from their onshore counterparts. On the other hand, some private organizations produced soundly reasoned estimates that looked beyond the resources themselves to the cost of their recovery and to the marketplace. The pot of gold did not look as glamorous in this cold light, and subsequent events have proved these pragmatic views largely correct.

As is always the case in such a public forum, a good deal of promoting took place. A scheme for mining diamonds just off the beaches of South Africa was actually implemented, but after some hard developmental work was closed down for lack of profits. An enterprising marine geologist sold gold shares off the coast of the island of Fiji, but has not been heard from recently. The U.S. Geological Survey wrote a detailed report on what they called "a bonanza in gold" on the Alaskan shelf near Nome, but to my knowledge that gold, if it exists, is still under water.

Actually, minerals in beach and nearshore sands have been exploited for decades. Offshore tin deposits in Indonesia and Thailand make a major contribution to the world supply, a significant amount of the metal titanium is obtained from Australian beaches, and nearshore iron sands are worked in Japan. The total value of such production is on the order of $50 million per year.

These resources, however, are mostly the exotics. Two quite different classes of marine mineral resources occur on the continental margin (again, ignoring petroleum) that have or probably will have more economic significance. The first of these is phosphate, and the other is building materials such as sand, gravel, and lime.

Phosphate, an essential fertilizer needed in all countries, is produced in just a few places: primarily the United States, Morocco, Tunisia, the Soviet Union, and four Pacific islands. Eight nations control the world supply of phosphate. The number of major consumers is far larger: Australia, Germany, Japan, and India are almost totally dependent on outside sources. For several decades we have known that nodule-shaped phosphates are common in many places on shallow offshore banks and plateaus and that some of these deposits, for example, off the coasts of California, New Zealand, Mexico, and South Africa, contain large tonnages of phosphate. The possibility of exploiting these offshore phosphate deposits began to be entertained seriously by scientists and some marine industries about 15 years ago, and a good deal of effort has since been invested in assays, resource estimates, and technology development. For many countries dependent on remote sources, such as India and Australia, the prospect of offshore phosphate reserves, for either immediate or future use, is quite attractive. Even the United States has some interest in a phosphate reserve within its own waters because, although we are now a net exporter of phosphate, this situation will change as existing reserves are depleted, and we will become partly dependent on imports as early as 1995.

As with so many potential marine mineral resources, the concept of mining offshore phosphates came to nought and has not been heard of for some years. This is not the result of a lack of phosphate deposits, which are plentiful and in some cases, such as California, are located favorably to markets. The undoing of the phosphate dream is economics. The grade of offshore phosphate reserves is invariably lower than that of land phosphates, and the lower grade, combined with higher extraction costs and some processing problems, renders the exploitation of offshore phosphates unprofitable in the present market. The price differential is small, just dimes per ton, but large enough to deter offshore phosphate mining until scarcity or the formation of a cartel of phosphate-producing nations drives up world prices or global transportation costs make local sources more attractive. Offshore phosphates are thus a resource of the future—a future probably not very far away.

Even less glamorous than phosphate fertilizer are sand and gravel, but they are equally essential if we continue to cover the entire face of the earth with concrete. In recent years, the annual consumption of concrete in the United States has been about four tons per person, a truly remarkable amount. Vital as sand and gravel are, however, we are not willing to pay a high price for these commodities. Because concrete is used in huge amounts and must not cost very much, the basic constituents—sand, gravel, and cement—cannot be obtained far from their use; transportation costs would be too high. Nature, unfortunately, has deposited sands and gravels in river terraces and flood plains, which also are the most favored locations for building. Thus the use of the land for producing sand and gravel competes with its use for building, and expanding cities swallow up the very ground that could supply the building materials for future construction.

As anyone can see there is a lot of sand in beaches, and some gravel too. Beach sand, however, is too well rounded and too smooth to make good concrete, and beach gravel is not really abundant. Moreover, although citizens demand houses and freeways, they would not like to see their beaches carted away for construction material. Happily, during the ice ages, when the level of the ocean was low, rivers traversed the newly emerged coastal plains and, swollen with the meltwaters of glaciers, laid down extensive deposits of sand and gravel of excellent quality. This happened mainly in northern latitudes of course and not in the tropics, but thus far the main population centers are also concentrated in those temperate zones.

The British, who began to experience a shortage of building materials about a quarter century ago, have realized for quite a while that a large resource existed on the continental shelf and accordingly have extracted a major share of their sand and gravel from the North Sea. Denmark and Holland have followed suit more recently. In the United States, Los Angeles on the western side of the continent and the megalopolis that extends from Boston to Norfolk, Virginia, in the east are rapidly running out of adequate sand and gravel supplies and have been eyeing the offshore deposits with some interest. The U.S. Geological Survey has demonstrated that ample deposits are available on the eastern continental shelf and smaller ones appear to exist along the California coast.

Why, then, is there not a flourishing offshore sand and gravel industry in the United States? The reasons lie both in the low value of the product and in the structure of the market. The price of sand and gravel at the building site, although locally variable, is about $8 per ton, most of which is transportation cost; production cost at the pit should not exceed $1 or $2 per ton. With modern marine dredging techniques, such a production price can certainly be met, but it would require a large capital investment in dredges and barges and a continuous, high-volume operation to make the investment pay off. Although it is comparatively cheap to idle an operation on land since few workers and not many bulldozers are involved, ships cost approximately the same whether they are working or not, and the investment capital for a marine operation runs from tens of millions to perhaps $100 million. Consequently, in order to be viable, an offshore operation must obtain a major share of the market. Alas, the sand and gravel industry traditionally has been very fragmented, with numerous small operators tied together in complicated and sometimes peculiar ways. In the Los Angeles area, the sand and gravel supply is controlled mainly by a small number of cement companies, whereas less respectable social elements have a hold on the industry in some eastern seaboard states. Accordingly it would be very difficult for a major offshore operation to penetrate the market and acquire a share large enough to justify the investment.

Inevitably these problems must eventually yield in the face of the rising shortage of construction materials, as will other impediments, such as environmental concerns. Dredging for sand and gravel—and for other offshore resources as well—is risky to the environment and potentially highly damaging. Many of the principal sand and gravel deposits are also the chief fishing grounds for shrimp and bottom fish. Furthermore, we do not know what the restorative capacity of the seafloor is, nor what will happen to the mud and debris that the currents may carry away from the dredge site to unknown destinations. I am inclined to predict that a compromise will eventually be worked out, but the first operator to apply for a dredging permit off North Carolina—or anywhere else—is bound to have a major legal battle on his hands. Already, the marine sand and gravel extracted annually is valued at half a billion dollars. Small as this may seem compared to $10 to $15 billion for fish and more than $40 billion for oil and gas, it is a respectable amount and certain to grow.

Sea Brines and Geostills

Until quite recently geologists regarded the oceans as kind of a cosmic garbage pail, a place where all the debris swept by wind and water from the continents ultimately comes to rest. The deep ocean has accordingly been regarded as of little interest to the economic geologist either as a source of fundamental knowledge or as a potential mineral resource.

New insights have drastically changed this attitude. Continental drift and plate tectonics show that the ocean is not an inert container, but a live and dynamic place where, at the mid-ocean ridges, new rock is continuously formed and heat generated, where volcanoes erupt and molten lava is intruded into the seafloor. These processes are quite analogous to those that are thought to form many of the ore bodies on land, and thus might do so in the ocean as well. At the very least, the study of these processes in the ocean should yield new insights that would help us in locating ore bodies on land.

We have long known that ores generally are deposited from hot, concentrated chemical solutions that rise through cracks and fissures in the earth's crust. Initially, because the ore bodies so often consist of metal sulfides, the solutions were thought to be rich in sulfur. In the last decade we have come to realize that fluids rich in chlorides are particularly effective in leaching metals from rocks, transporting the solutions, and depositing the metals elsewhere as ore veins or ore bodies. Since seawater is nothing more than a strong solution of sodium chloride, hot seawater can be regarded as a potentially excellent carrier for ore-forming substances. The ocean thus provides both an inexhaustible supply of brine and the processes necessary to leach metals, transport the solutions, and deposit ores.

How the system might work can be seen in Figure 11 (next page). On the crest of a mid-ocean ridge, hot lava is intruded into the crust and extruded upon the seafloor. As the lava cools to form the normal oceanic rock, basalt, it contracts and numerous fissures and fractures open. Through these fissures, seawater penetrates into the newly formed basalt and the molten lava underneath. As the seawater percolates, it is heated and thus becomes less dense, so that a circulation system is set up whereby cold seawater flows down into the cooler fissures on the flanks of an eruptive zone and hot water rises in the center of the volcanic activity. Where the hot brine comes into contact with lava or hot rock, it extracts soluble elements, such as iron, manganese, copper, nickel, lead, and cobalt, and carries them to the surface. The seawater itself is also rich in various elements, and some of these, such as potassium, sodium, and magnesium, are taken up by the hot rock with which the seawater comes in contact. The exchange process

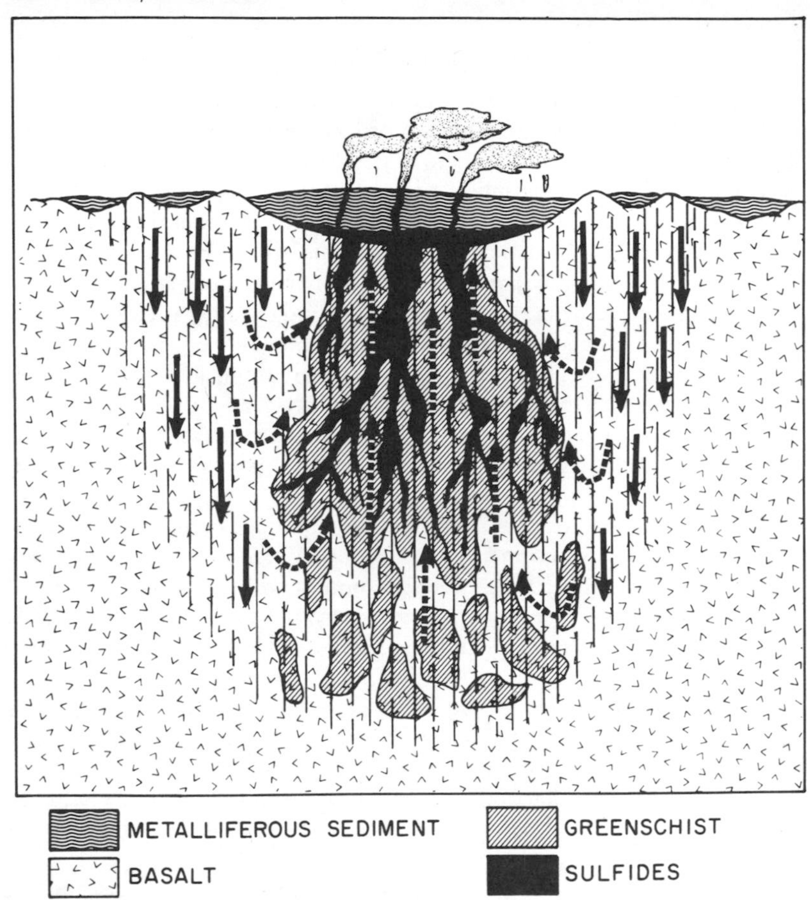

Figure 11: The interaction of seawater and seafloor spreading on a mid-ocean ridge alters the original volcanic basalt to greenschist, and causes iron sulfide ores to precipitate within the rock and metal-rich muds to deposit on the seafloor. Arrows show the circulation of seawater through fissures and cracks down on the cool margins of the volcanic body and up in the hot center. Drawing is taken from John B. Corliss, "The Sea As Alchemist," *Oceanus,* Woods Hole Oceanographic Institution (vol. 17, 1973-74, p. 39).

with seawater transforms the original basalt into quite a different rock type called *greenschist.*

Not all of the metals dissolved in the brine are carried to the seafloor. One of the most abundant metals in solution is iron, which is oxidized in seawater and combines with sulfur, also present in large quantities in seawater, to form sulfide ore bodies that also contain copper, silver, and lead. The rest of the dissolved metals make their way to the seafloor in the column of hot brine rising through fractures and fissures. At the seafloor they come in contact with the very cold ocean water and precipitate as a reddish-brown mud consisting mainly of iron and manganese hydroxides, with copper, cobalt, and nickel as additional components as well as some minor quantities of metals, such as zinc, scavenged from seawater.

These metal-rich muds are very common on mid-ocean ridge crests where they rest immediately on top of the basalt. The *Glomar Challenger* has drilled into such muds on the flanks of mid-ocean ridges in the Atlantic and Pacific underneath a later cover of oceanic sediments. Metal-rich muds, ranging in thickness from a few feet to tens of feet, are thus widespread oceanic deposits. The dominant components, iron and manganese, are not of much value themselves because there are such large deposits of these metals available on land which can be exploited more cheaply. The copper, nickel, zinc, and other components, however, although present in only a few percent, are scarce on land and are usually obtained from ores not much richer than the oceanic muds; consequently, the deep-sea, metal-rich muds might eventually become a usable mineral resource competitive with ore bodies on land. Mining the muds would probably not be too difficult because they are very watery and could be made into a slurry and pumped to the surface of the sea for processing.

At this point we have a deposit of metal-rich mud at the ocean floor, resting on basalts. Along fractures and fissures, the basalt has been transformed into greenschists in which the rising hot water has formed veins of sulfide ores of iron, nickel, and copper. On the mid-ocean ridges we can surmise the presence of the deeper parts of this complex (see Figure 11), but cannot confirm it because there are no practical means for penetrating deep into the oceanic crust to locate the ore bodies. However, similar sequences of rocks, complete with metal-rich muds on top and sulfide ore bodies within, are quite common on the continents where they appear to have been placed by collisions of oceanic and continental plate edges. One of the world's oldest copper mines is in such a slice of an ancient ocean floor on the island of Cyprus (the name itself is the Greek word for copper). This copper was mined, probably as early as 3000 to 4000 B.C., from rocks that resemble rather precisely the diagram of Figure 11, complete with internal sulfide ores and the metal-rich muds on top, which the Greeks called ochres (iron-rich) and umbers (manganese-rich) and which were used for paints and cosmetics.

The story does not end here. We have seen in Chapter One that the inevitable result of creating new crust on a mid-ocean ridge is that an equivalent amount of old crust must be destroyed somewhere else. The elimination of excess material takes place in collision zones where the oceanic crust of one plate descends below either a continent or another oceanic plate. As the crust goes down, it takes its load of sediments, metal-rich muds, basalts, greenschists, and associated sulfide ores, as well as a lot of seawater, back down to the depths from which many of the constituents, a long time ago and in a different place, originally rose to the surface. Some of the soft upper crust may be scraped off against the overriding plate, but a significant portion probably finds its way to great depths where the pressure and temperature are high. The sediments, basalts, and greenschists, having been altered by seawater at relatively low temperature and pressure, are not stable under the conditions at great depths. Consequently, they melt sooner and more easily than the rocks underneath and above them which are at home at those depths. The water-rich melt will start to rise wherever it can find a crack or fissure. The diagram of Figure 12 shows how we envisage this process.

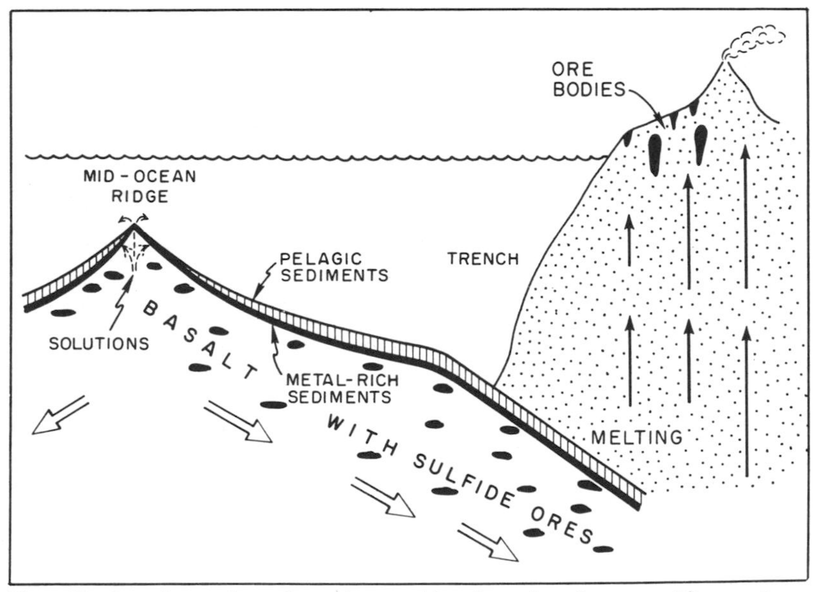

Figure 12: Oceanic crust formed at mid-ocean ridges descends under an overriding continent at a collision plate boundary. The oceanic crust carries metal-rich muds and sulfide ores with it to great depths where the sediments and ores melt. The metals are leached by water in the sediments and carried upward by the brines. The brines precipitate copper ores near the surface in the mountains of the overriding plate, where subsequent erosion makes them accessible to mining.

The molten material, also containing the constituents of the metal-rich muds and the sulfide ores, will gradually melt its way through the overlying plate. As the molten mass rises, the seawater, being the most volatile part, segregates from the mass, leaches copper and some other metals from the molten lava, and carries these metals in solution to shallow depths below the surface where they will precipitate as copper ores known as *porphyry coppers.* Porphyry coppers form the world's largest copper resources and are nearly always associated with present or former colliding plate boundaries. The copper ores of Chile lie just behind the collision line of the South American and Pacific plates. The large copper deposits of Arizona, Nevada, and Utah can probably be explained as the result of a collision between the North American plate and a plate once located in the eastern Pacific but now completely consumed by the collision.

Together ocean water and seafloor spreading provide us with a good deal of improved (although in part not proven) insight into the processes that make ore bodies. In addition, they may furnish us with some metal deposits that will be useful in the future, when other resources have become scarcer and hence more expensive than they are now. The oceanic mineral deposits most likely to be exploited soon, however, are not the metal-rich muds and the sulfide ores, but the *manganese nodules,* similar in composition but wholly different in origin and distribution.

Harvesting the Manganese Nodule

Millions of square miles of the Pacific Ocean floor are paved with brownish-black nodules shaped somewhat like potatoes and ranging from one-half inch to several inches in diameter. The nodules consist primarily of hydroxides of manganese and iron. They rest as a single and often closely packed layer on the sediments of the ocean floor. The nodules are found predominantly in the deeper parts of the oceans, well away from regions of rapid sedimentation near the continents, outside the equatorial zone of calcareous sediments, and far from the crests of mid-ocean ridges. Figure 13 (next page) gives a good idea of the enormous area occupied in the Pacific by these peculiar formations. If one takes into account that they also occur extensively in the Indian Ocean and to a lesser extent in the Atlantic, it is evident that they are a very ubiquitous type of sedimentary formation and represent an enormous volume of manganese and iron.

We have known about the existence of manganese nodules since that granddaddy of oceanographic expeditions, the *Challenger* expedition of 1872–76, found the first nodules. They have been studied extensively, and we now have reasonably good regional maps that show their gross distribution, even though on a smaller scale there is still much to be learned. Different as they may be in occurrence, shape, and consistency from metal-rich muds, chemically they are close cousins. The similarity extends even to minor constituents. In addition to an average 6 percent iron and 20 to 25 percent manganese, the nodules contain on the average about 1 percent each of copper and nickel and somewhat less cobalt. The concentrations of the minor elements vary a good deal from region to region, but we now have fairly accurate maps that show what amounts of copper, nickel, and cobalt can be expected in specific parts of each ocean. There seems to be some relationship between composition and environment—especially water depth—but its precise nature is not yet understood. Recently, manganese nodules have also been found in shallow waters—on the Blake Plateau off eastern Florida and even in Norwegian fjords and Scottish lochs. These shallow nodules tend to consist almost entirely of manganese and iron and lack minor components.

Manganese nodules are intriguing not only because of their distribution and their chemical variability. We do not know, for example, where the metals come from. The similarity between the nodules and the metal-rich muds of the mid-ocean ridge crests suggests a volcanic origin for both, but the nodules preferentially occur in the deepest and oldest parts of the

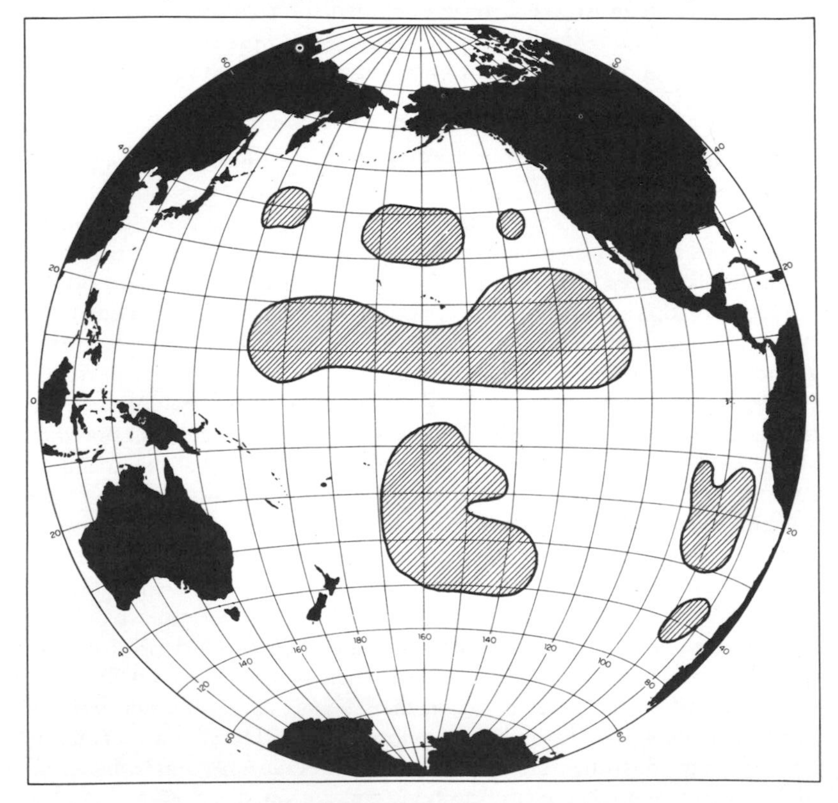

Figure 13: Areas with the highest concentration of manganese nodules in the Pacific Ocean. Average concentration in shaded regions is about 20 pounds per square yard with an average content of copper, cobalt, and nickel combined of 2 percent.

ocean, far away from active volcanic centers and mid-ocean ridges. It therefore seems more likely that the nodules acquire their constituents by extraction from seawater or from the sediments on which they rest. We also do not know how and why they form.

The evidence suggests that they grow very, very slowly—at a rate of perhaps only one millimeter (.05 inches) per million years. Do they grow as a result of purely chemical processes? Or is there some biological, perhaps bacterial, mechanism involved? This last possibility has recently found some vigorous supporters but is not widely accepted. If they grow so slowly, how do they manage to stay on top of sediments that accumulate from ten to a thousand times faster? Although some fancy mechanisms have been suggested—such as burrowing worms carrying them continuously upward—nobody really knows the answer. Perhaps they form faster than we assume, but episodically, and then become buried. We might, after all, have overlooked layers of nodules buried deep in the sediments. If that is true, however, it would mean that we are now experiencing an unusual period of nodule formation, and it is never very satisfying to call on an explanation that requires the present to be a very special time.

A good deal of effort has gone into solving these and other mysteries of the manganese nodules; there are numerous active manganese nodule specialists and even a major manganese nodule project supported by the National Science Foundation. But we still have no real answers to most of the basic questions raised by the nodules, and the principal advance of our knowledge since the *Challenger* expedition is a better description of their geographic distribution.

There is quite a bit of irony in the current state of affairs, because the nodules, whose very existence was established by pure science, are now the most promising oceanic target for exploitable minerals, a target that is plagued not by these scientific problems, but by quite a different set of difficulties that are, in the main, not scientific at all.

Since the late 1950s and early '60s the manganese nodule deposits have attracted much attention as a potential marine mineral resource. The interest is not in the manganese and iron; these metals are plentiful on land and so widespread that no potential monopoly positions could put the squeeze on either price or supply. That, however, is not the case for the minor elements in the nodules. Copper, nickel, and cobalt are either no longer available in abundance, are concentrated in a few countries or regions, or are already being mined from ores so low in grade that they are hardly more economical than the nodules.

The early enthusiasm for manganese nodule mining was remarkable more for its optimism than for its hard-nosed approach to the technological and economic questions. Many a small operator found to his dismay that knowing of a vast mineral deposit is not the same as exploiting it for a profit. The deposits are vast: in the Pacific alone, a recent estimate suggests the presence of about 350 billion tons of manganese, 200 billion tons of iron, 15 billion tons of nickel, 8 billion tons of copper, and 5 billion tons of cobalt. Smaller but nonetheless sizable amounts occur in the Indian and Atlantic oceans. Moreover, there is no secret about where this hoard is located. By contrast, in the petroleum and most mining industries, just *finding* the deposit tends to be at least half the game, and the accomplishments of the industry are measured in large degree by its success in exploration. The fact that oil and most ores are *not* there for everyone to see simplifies the technical and legal problems of establishing ownership and acts, to some significant extent, as an industry stabilizer.

Finding manganese nodules is easy: anyone can sit down with a map and mark off the square he likes. Certainly, for efficient exploitation more work is needed, because their distribution on the ocean floor is not all that even and the composition not quite as uniform as the maps indicate. In principle, however, the problem is neither finding the resource nor matching the rate of discovery with the rate of consumption and that of exhaustion of reserves. Thus, different modes of regulating and stabilizing production, investment, and profits must be developed.

Easy as the nodules are to find, their recovery is not for everyone. Initially, the capital and technology required for economic exploitation were vastly underrated, but by some ten years ago mining experts began to recognize that the undertaking would require very large investments, say $100 million, and very long lead times of a decade or more. Only a large, technologically advanced industry is capable of this sort of enterprise. The estimates have certainly not proved to be excessive, and as a more realistic attitude began to prevail, the number of potential entrepreneurs diminished until there are now perhaps only four or five major groups with a strong stake in the field, including companies from the United States, Canada, Germany, and Japan.

The problems involved in a successful manganese nodule mining effort are several and dwarf the relatively simple task of exploration and resource evaluation. First there is the question of what technology is needed to recover the nodules and bring them to the surface. Two competing techniques are under consideration at the present time: dredging with a continuous bucket chain, and using a modified form of airlift suction pump. Neither of these concepts is easily executed in water depths of 15,000 feet (4,000 meters) or more. In addition, to recover the nodules efficiently it is necessary to make a clean sweep of the seafloor, and that implies sophisticated control and inspection devices on the bottom.

The next problem is what to do with the nodules once they are safely on board. The metals of value make up only about 2 percent of the bulk, and the good mining sites are located thousands of miles offshore. At the present it appears that the only feasible course is to transport the raw nodules to some onshore refinery, but the cost is high and transportation becomes one of the big items in the operations budget. Moreover, maintaining a steady flow of shipping necessary for a large volume operation at sea—and no other will ever pay off—and with variable weather conditions is a formidable logistics problem. Refining the nodules at sea would obviously provide a major economy, but this is likely to be a difficult process requiring a large plant and much energy. There are some fancy schemes for operating a nuclear-fueled refinery at sea or for using thermal ocean energy, or wave or tidal energy, but they are, as yet, far removed from reality. Nevertheless, it is fascinating to contemplate a completely self-contained mining and processing plant floating in the middle of the Pacific. Obviously that would open entirely new vistas on security, property rights, and even territorial rights.

Problems associated with the extraction of copper, nickel, and cobalt from the nodules have also not been solved. Because these elements are so intimately contained in the basic iron and manganese hydroxide structure, the required technology is entirely different from that used for conventional ores of these metals. Leaching the metals with our familiar chloride brine holds the most promise, but pilot plant operations have barely started.

The future of manganese nodule exploitation thus hinges on technology and money rather than on geology and oceanography. The technical and financial aspects are especially critical because present estimates suggest that there is no easy money to be made. With a yield of 20 pounds of nodules per square yard of the ocean floor and with nodules containing 2 percent of copper, nickel, and cobalt combined, a marginally profitable operation appears to be possible. The estimate, however, includes many uncertainties. No large-scale recovery experiment has yet been undertaken, hence we do not really know what the cost per ton will be in a large mining operation encountering, no doubt, somewhat variable ocean floor conditions. The transportation cost, a very large factor, fluctuates with the world shipping market, which itself has never shown even the slightest tendency to stability. Today, we can only make crude estimates of the cost of refining the metals. Worst of all, the world price of copper, nickel, and cobalt—against which the profitability has to be measured—has been far from steady in recent years.

In fact, no serious scarcity of any of these three commodities is expected in the next quarter century, and currently there is even a glut of nickel because of recent developments of large new deposits on land. Although copper, nickel, and cobalt are essential in modern industrial society, the demand for them is not particularly large. If marine metals could corner 10 percent of the world market, which would be an exceptionally large share, the annual gross of marine mining would be about one billion dollars, not so large an amount in view of the enormous risks and the size of the required investment. Why, then, should we bother with the marine deposits at all?

The answer to this question is political rather than economic. In 1975, the United States imported about 70 percent of its nickel, primarily from Canada and Norway; nearly all of its manganese from Brazil, Gabon, South Africa, and Australia; and all of its cobalt, mainly from Africa. Over the next decade, the volume of our imports will rise dramatically and our dependence on outside sources will become almost total. Most of the sources are either in politically unstable areas, potentially vulnerable to control by cartels, or connected with the United States by long, easily intercepted transportation lines. Japan, Germany, and many other major industrialized countries are far worse off than we are. If wisely handled and properly regulated, a marine manganese nodule operation could have a stabilizing influence on world prices, guarantee to some industrial countries supplies that would be less affected by global strategic and political considerations, and develop a major resource for future world use.

It is understandable that such a prospect does not fill the present suppliers of these commodities with undivided joy. Furthermore, for a variety of reasons, not all of them sound, many of the developing countries have exceedingly high expectations of the wealth that can be obtained from oceanic mineral deposits. These expectations generally are not justified by the realities of the resource and its exploitation, but they are nonetheless a powerful political force. To these nations, particularly the so-called Fourth World (the destitute have-not nations of the world), the exploitation of resources in the open ocean *by* and *for* the benefit of the rich industrialized nations is pure anathema. The exploitation of marine manganese nodules has, inadvertently, become a key issue in the attempts of the United States and other countries to establish a new code of laws for the sea. It is rare that the activities of oceanographers produce a major political and historical event, but when this happens they are, as we shall see, among the first to suffer the consequences.

Who Owns the Sea?

Most of us believe that the sea historically has been free for all to use and traverse. In principle this is a noble concept, but in practice those few who have the skills and the means benefit more from common property than those less endowed. The concept of a territory that is common property is not new. One of the older social institutions in northwestern Europe, amply predating the introduction of Roman law, was the recognition of common land belonging to all that cannot and should not ever be appropriated by any individual or limited group. The communal grazing grounds and woods that formed part of most European farming villages from prehistoric to medieval times provide the best example of such a *commons*. Explicitly or implicitly, most of us regard the ocean as the last holdover of this principle, limited only by the narrowly defined territorial interests of coastal nations. In this light, the tendency in recent decades to extend national control farther out to sea strikes us as the beginning of the dissolution of an ancient and valuable custom.

At the root of the collapse of the commons lies man's inability to reconcile his own profit with the common good. The issue is basically simple. Think of ten farmers, each grazing ten sheep in the common pasture. If a single herdsman adds one sheep, he will increase his profit by 10 percent while diminishing the food supply on the common land by only 1 percent. To him, this appears a harmless thing to do. On the other hand, if the commons requires maintenance, all ten farmers must be persuaded to put in a share of their time for which the visible profit for each individual will be small; this, generally, is a delicate and difficult undertaking. The fishing industry beautifully illustrates the problems of the commons of the ocean. It is to the advantage of each fishing vessel to catch as much as it can, while it is difficult or impossible to convince individual fishermen, villages, and nations that they must agree on limits necessary to preserve the stock. In the long run, man has not been very successful at cooperative enterprises that restrain the individual advantage. Attempts at management of fish stocks have come late (in the late 19th century), have had a spotty record, and generally have not been able to stem the rapid decline of many once-vast fish populations. This failure at large-scale cooperation generally leads to the breaking up and parceling out of the commons, on land during the late Middle Ages, at sea beginning now. The complaints of the individual herdsmen who lose in the draw are being heard everywhere.

All of this makes a nice and sensible story, but the sea was not regarded as a commons until quite recently. The concept of the freedom of the sea has a short history; it was first explicitly stated by a young and later very famous international lawyer from Holland, Hugo de Groot (Grotius). In 1605 he wrote a pamphlet, *Mare Liberum,* which was not an attempt to codify an existing tradition but rather was a response by the small Dutch nation to threats against their far-flung maritime trade by territorial claims of the British. (If you are not strong it helps to have the law on your side, even if you have to write the law yourself.) England, on the other hand, was

Mild Peril at Sea by Paul Klee.

following old tradition by extending her claims of territorial rights farther and farther out to sea. This tradition can be traced deep into the Middle Ages, when several nations of northwestern Europe established trade and fishing monopolies at sea and protected them with treaties and the force of arms. Later, the levying of customs duties on shipping through straits and claimed territorial waters further extended territorial rights at sea.

Gradually, however, the number of powerful seafaring nations and the global extent of their interests increased, and the defense of monopolies was replaced in the 18th and 19th centuries by a strong stake in the right to go wherever you want. A free ocean is necessary to expand a nation's trade, to maintain its overseas territories, and to exercise the power to protect these interests. The concept of the freedom of the seas was well suited to those expansionist times and accordingly prevailed until the end of World War II. In fact, most naval wars were fought not to protect territory, but to open or maintain free passage. There are no problems, really, with a commons if all you want to do is walk across it. Until the middle of this century the principal restriction of the freedom of the sea concerned coastal defense and control of smuggling, and a narrow territorial zone satisfies both. Based on an assumed range of coastal batteries of three miles (somewhat optimistically, because 18th-century cannons did not reach much beyond one mile), this territorial limit was established firmly and universally.

After World War II many of the former colonies of the Western maritime powers became independent, and thereby fundamentally changed the basic realities that underlie the concept of the freedom of the sea. The lesson was slow to sink in, however, because an international conference in 1958 again reaffirmed the freedom of the sea and only tinkered with territorial limits. The treaty, however, added one new principle which resulted from the growing recognition that there are marine resources other than fish. It defined, quite artificially, the continental shelf as a zone limited by the 200-meter depth contour and assigned all rights to the resources on and under the seabed of the shelf to the adjacent coastal nation. The legal status of the water above the shelf was not changed. The chosen limit was quite unnatural and unfortunately did not correspond to geological and resource realities; it has thus given rise to a protracted and unresolved debate between geologists and lawyers.

Entirely apart from this international convention stands a series of bilateral and multilateral treaties that were designed to regulate fishing and preserve fish stocks. These treaties, which began in the late 19th century, bind only the partners—others are free to do as they please. As a result, the agreements tend to have very limited effect, as is demonstrated by the International Whaling Convention: the biggest whaling nations, Japan and the Soviet Union, never joined and thereby defeated all attempts at conservation of the whale herds.

The success of fishing treaties was so questionable that a number of nations highly dependent on fisheries began to think in terms of unilateral action by expanding their sovereignty over entire fishing zones. Such actions sprang from two basic motives: a laudable attempt to adapt fishing practice to what the ocean can bear, and a selfish desire to keep the fish for one's own nationals and lock others out. Extended fishery zones out to 200 nautical miles from the coast are claimed, for example, by Peru and Ecuador, both highly dependent on offshore fishing. However, as a management device, unilateral action has not been very successful, because most pelagic fish range over vast reaches of the ocean during their life cycle. The territorial claims have been the source of much indignation among U.S. tuna fishermen, but in the uproar it has sometimes been forgotten that in 1966 the U.S. itself established, quite unilaterally, a fisheries monopoly out to the 12-mile limit and on March 1, 1977, extended it to a 200-nautical-mile limit. Many other nations have established a great variety of limits, which usually are not internationally recognized. Such unilateral limits give rise to much friction and unpleasantness, as evidenced by the codfish war between England and Iceland.

As a result, by the early 1970s a patchwork of rules and claims had developed that made very little sense, while everyone was struggling with the real meaning of the resources of the seabed on the shelf. The situation became progressively more emotional as it was recognized that the seabed resources really do exist and possibly even extend to the deepest ocean, a part of the world that no one had given much thought to before. The United States has traditionally favored as much freedom of the sea as could be maintained internationally and under the clamoring for special rights by its own citizens. This attitude, a rational one for a major sea power with far-flung economic interests, dates back to Thomas Jefferson's very narrow definition of a neutral zone of waters along the coasts of the new nation.

Several years ago the United Nations conscientiously, but as it turned out somewhat optimistically, decided that it was time to resolve these matters and called the first of a series of international conferences on the law of the sea. Since the first conference in Caracas in 1974, two more have been held and another one is scheduled for 1977, but the issues are far from being resolved. The reasons for the difficulties are complex. First, there is the issue of the freedom of the sea as a stage for the application of political and military power in the protection and preservation of national interests, mainly trade. This particular issue figures strongly in the U.S. position and brings us partners such as the Soviet Union, France, and Great Britain. Most of the Third World nations, understandably, have little sympathy with this rationale for the freedom of the sea, which they regard as a stubborn residue of imperialism and colonialism. Then there is the question of the resource zones, the old much-claimed 200-mile limit for fisheries control, now broadened to include all other real and imagined resources of the sea. Next follows the question of the ownership of the resources of the open ocean, especially minerals, which occur beyond reasonable national limits.

As if these problems were not themselves sufficiently confusing and conflicting, an entirely new group of participants has joined the dialogue. Whereas in the past, discussions of the law of the sea mainly involved maritime nations, in particular those actually capable of utilizing the sea and defending their interests, the new dialogue includes virtually the entire world. In this enlarged forum, the majority of the nations either do not even have coastal borders or, if they do, they have never been involved to any significant degree in maritime activities.

Among this new class of non-maritime participants, loosely known as the Group of 77 (although they number more like 100), the notion has taken hold that the high seas are a traditional commons with very rich resources that belong to all. These nations, following the lead of Malta, see it as in their interest to minimize the extent of all national limits and maximize the area, but not the freedom, of the high seas. This group, which tends to be short on real expertise regarding the sea but entertains very optimistic views on the value of ocean resources, has been claiming the riches of the sea as the birthright of all mankind. One can sympathize with this point without losing sight of the fact that the reality is both more complicated and much less profitable than they perceive it.

The debate is further complicated by the highly divergent interests of the members of the Group of 77. While some among them desire immediate development of the expected treasures, others have their own domestic resources (with which the marine minerals would compete) and thus wish to retard development of the oceanic deposits as much as possible. Virtually without exception, the nations in the Group of 77 possess neither the capital and know-how to develop the oceanic resources nor the power to regulate and police the production for the common good. Their proposed scheme of an international authority to license and supervise resource exploration and exploitation and to receive royalties is flawed by an almost punitive attitude toward the United States, one of the few nations actually capable of recovering the treasure. In general, these nations disapprove of the capitalist approach, yet they will have to live with the fact that today the required skill and capital are available only in the Western industrial nations. There is also a widespread lack of realistic appreciation of the costs, technologies, and rewards involved.

With such fundamental divisions and conflicts of interest on so many fronts, it seems likely that the Conference on the Law of the Sea will accomplish little of substance. Inevitably, this means that, as far as the market will bear, each nation will claim national resource zones which include complete sovereignty of the coastal nation over fish and mineral resources. This unilateral approach will fragment the management and exploitation of marine resources and will sharply reduce the area of the free ocean, perhaps by half. It is not clear that the conference will be able to resolve even this problem, and if it does not, unilateral claims like the recent establishment of a 200-mile zone around the United States are inevitable. I do not believe that any agreement at all will be achieved concerning the high seas because there seems to be no common ground between the various parties that rests on solid considerations of economics and power. Hence, the high seas will probably remain free, although much diminished in size, and open to exploitation by those few nations who have the capital, the technology, and the need for the resource. These nations will face the entirely new problem of protecting their property rights and investments in an international no-man's land. The marine industry, dead-set against international regulation, seems to be confident that these problems can be solved to its satisfaction and profit.

In all this high-level commotion, ocean scientists have been running around like so many scared rabbits. Oceanic phenomena pay little attention to national boundaries, and few scientific problems in the ocean can be resolved without a global or at least ocean-wide approach. With the encroachment of territorial limits on the high seas, the freedom of motion of scientists is being daily diminished to the point where some feel that the future of the science and hence the benefits to mankind that can be expected are in serious jeopardy.

Many nations are suspicious of the activities of marine scientists: too often, military reconnaissance and industrial exploration have masqueraded as academic enterprises. Events like the capture of the *Pueblo* or the unmasking of the manganese nodule ship *Glomar Explorer* as a front for a C.I.A.-sponsored intelligence effort are not easily forgotten. In fact, the latter event has impeded the access of the purely scientific drilling vessel *Glomar Challenger* to some parts of the oceans. Scientists themselves have not always been careful to obey the rules of the host nations or to include their foreign colleagues in their own activities and give them the benefits of their results. In their attempts to secure access to foreign territorial waters, scientists have not been well supported by the State Department which, although paying lip service to the principle of scientific freedom of the seas, has tended to regard the interests of ocean science as trump cards to trade for other concessions.

As a result, many academic oceanographic institutions have chosen to build direct bilateral relationships with the countries whose waters are of interest to them. Quite often, this involves more than mere arrangements for collaboration and requires that considerable financial assistance be given to the country involved to help train the scientific personnel and build the organizations and apparatus that will enable the country to participate in a meaningful way. The oceanographic institution at Oregon State University, for example, has built an elaborate Latin American cooperative program, while the University of Miami specializes in the Caribbean. Basically, of course, such arrangements are a form of foreign aid, but under the mysterious ground rules that prevail in Washington, most of the costly activities are funded out of the nation's research budget.

For most of us the great variety of talents upon which an ocean scientist must draw is one of the major charms of our profession. Human relations, management, fund raising, scientific work in physics, biology, and chemistry, technological skills, logistics planning, and ship handling make for a fascinating array of things to do. Now to all of this has been added nimbleness in diplomacy and international relations. As we stumble along, our principal comfort is that many of the other professionals do not appear to do much better.

U.S. submersible *Alvin*.

CHAPTER SEVEN

ATTENDING MARVELS

In the small hours of the third watch, when stars that shone out in the first dusk of evening had gone down to their setting, a giant wind blew from heaven, and clouds driven by Zeus shrouded land and sea in a night of storm.

Homer, *The Odyssey*, Book XII

FORTY YEARS AGO, the distinguished paleontologist George Gaylord Simpson evoked the fascinations of the life of one naturalist in a book bearing the title I have chosen for this chapter. I cannot possibly do as well as he, but in this final chapter I feel compelled to speak of one special reward among the many that oceanographers enjoy. Most of our pleasures we have in common with the practitioners of other sciences: the curiosity about unsolved problems; the joy over an elegant experiment, a clean solution, or a meaningful set of data; the pleasure of seeing one's work in print, looking so definitive; the satisfaction of accomplishing something that might be useful to society and one's fellow man; the approval of one's peers; and, above all, the privilege of making a living out of a hobby. To me, none of these compares to the reward I wish to speak of in this chapter—the love of the sea.

Instead of attempting a poetic description from my desk chair, I shall rely on impressions excerpted from two of my cruise diaries, with as little editing as sloppy grammar and careless writing will tolerate. The first of these is from an ordinary cruise, one of the hundreds that have gone out to provide yet a little more data on yet another piece of ocean. The cruise took place on the research vessel *Atlantis-II,* which sailed in 1966 to study in some detail the Vema fracture zone in the southern North Atlantic. We left from Belem with a varied load of scientists and miscellaneous programs—including fishing for plankton, tracking currents with radio buoys, and sampling for radioactive fallout in ocean sediments and water.

The second cruise was far from ordinary. It took place in the summer of 1974 and accomplished the first visits by man in a research submersible to the divergent plate edge on the Mid-Atlantic Ridge to see firsthand the volcanic and structural processes involved in the rupture. It was an international enterprise: there were two French submarines, the *Cyana* and the *Archimede,* with their motherships the *Noroit* and the *Marcel LeBihan;* one U.S. submersible, the *Alvin,* with mothership *Lulu;* and a large research vessel, the *Knorr* of Woods Hole Oceanographic Institution, which served as the American laboratory ship and headquarters. Nobody had ever before visited these depths of the ocean in such a systematic manner, virtually a geologic mapping project, and we became exposed to a good deal of publicity, as well as attention from the press and high French and U.S. officials. The management of this large and cumbersome enterprise was wholly new to most of us who were accustomed to a rather independent and lonely operation with our own crew in mid-ocean. The cruise began and ended in the Azores.

Atlantis-II Cruise 20, April–May 1966, Belem, Brazil, to St. Thomas, Virgin Islands

April 1, 1966: The night after flying in was short and hot, and at six I cannot sleep and decide to go find the ship. After substantial confusion regarding its location and a search of 12 piers, we find her tied to a weed patch near some grain elevators, just in time for breakfast. I always admire the clutter of dirty linens, sea bags, and suitcases that marks the arrival and departure of scientific staffs.

April 3: The two days have been spent mainly in waiting and talking while captain, chief scientist, and agent (the last one not too actively, it seems) attempt to get the ship fueled. Because I woke up late each day and it rained hard in the afternoons, I have seen little of the city, but that little is favorable. The streets are clean and the people well washed, friendly, and healthy-looking. Surprisingly, in this land of mud and water, the streets are paved with cobblestones. River traffic is heavy from small canoes to handsome, rakish-looking sailboats and large paddlewheel steamers—all loaded with great numbers of smiling men, women, and children selling Brazil nuts. The buyer catches them and drops the estimated price. Hilarity when he overestimates, great wailing and picturesque despair when it is less than expected.

Yesterday night after a depressing cocktail party at the consul's, we went for drinks at the local upper-class house, said to be quite a spectacle. It was indeed: a large, darkish hollow room echoing to a small but loud band, ringed with tables. The outer set was occupied by numerous quite beautiful, well-dressed and groomed girls whose most outstanding characteristic was utter passivity. Not even by smiling straight at them could one elicit a reaction, even a blink; a mystery how they manage to lure their clients. They do not dance, they do not drink, they only speak Portuguese, and, presumably, they go upstairs. A very antiseptic place.

April 4: We find that the agent has not made any fueling arrangements. Probably it will be at least four days before our turn comes. We work up an alternative schedule including fueling in Paramaribo and arrange for a telegram. Find later that the captain gave the telegram to the agent for dispatch. He must be an innocent at heart.

April 5: The agent forgot to send the telegram and we are still waiting. The radio buoy out at sea hasn't been heard of for a day, but I suspect that Gary isn't listening for fear he may not hear anything. Rain for the past 36 hours; even the inhabitants of Belem cannot remember such weather and it keeps us locked up.

April 6: The consul has saved us; today we shall fuel at 0900 ahead of another vessel with a higher priority but a less influential consul.

April 7: Well, at sea again; there is a funny disorderly swell on an oily sea and my back takes poorly to the ship's motion, which is as bad as that of the *Washington*. The chief scientist decided to go home again, and we are now led by his assistant who appears depressed by the prospect and by the motion in the chief scientist's room which is up forward. I am sure he will do well but be seasick. Around 1000 hrs. we get rid of the pilot who, having almost been squashed between the ship and his tender, disappeared in volleys of unmistakable language toward Salinas, barely visible in a rain squall. The sea is slick, black squalls are all around with an occasional patch of lovely tender blue and white. Gary's drogue is still alive, beeping some hundred miles away. Everybody is acquiring his sea legs in his own way—mine is by sleeping. We train the new echosounder watch. Their reactions vary between such excesses of timidity in the face of this formidable instrument that they won't even read the chart and such rapid flipping of random controls before we can even yell "No" that we spend a chunk of the afternoon repairing the damage. After a dark and threatening sunset in the squalls, we arrive on our first station in the usual utter confusion, as if none of us had ever been to sea before.

April 9: Two days at sea, closed with a beautiful evening, warm, light breeze and the sun setting in black, gray, purple, and orange clouds against the tender yellow evening sky. Such evenings almost justify the effort of going to sea. Gary's drogue is blinking its light a mile away and we are waiting for a star to find out where we are. Supper was Boston baked beans, Woods Hole Saturday traditional, but not my favorite.

April 10: Easter Sunday, sunny and lovely morning. I spent an hour sunning myself on the fantail, but notwithstanding the colored eggs that were provided at breakfast, I did not have a real Easter feeling and find it difficult to imagine Easter egg hunts at home. Otherwise, the day wore on in boredom, more plankton tows, more bottle casts, more midwater trawls, and still we are within 36 miles from where we were four days ago. Even the sea surface is beginning to look familiar, and the rise and fall of the ship is faintly dreary. Late in the afternoon a flight of Leach's petrels brightened me up—lovely small birds chasing close to the water and right under the counter. I fed them some bread which they happily accepted, although how they should know it is edible is a mystery.

April 11: Still raining hard and there is short choppy sea. The A-II has a short, rough pitch and a strange roll because of the antiroll tanks. It is an almost random motion, short but unpredictable, and now I know how an ion buffeted by Brownian motion must feel. No stars since yesterday and our position is getting rather tenuous. The main things accomplished are a boat drill and my laundry, not too much for a whole day. The echosounder keeps troubling us; Vic devotes much time to it and it is frightening that this fragile thing holds the key to my success or failure.

April 12: Our first piston core was a useful exercise allowing us to get a lot of confusion and yelling and shouting out of our system. That was all it did; no core after seven hours because a little bolt stuck 1/16th inch too far out. Looking up from the crowd of sweating, milling, half-naked, cursing men on the fantail, I noticed a row of deck chairs on the boat deck occupied by cool-looking ladies in fresh clothes and sun hats, holding tall cold drinks. There is merit to having women on board.

April 16: Many busy and long days with closely spaced cores. The operation now goes well, but we are unlucky and get little penetration because of hard layers. George does not like the Woods Hole gear, and his language has sunk to a level where I have had to caution him. Now he constantly looks over his shoulder to make sure there are no women within earshot. I don't think they would care, they are a hardworking bunch and good members of the crew. Tonight inspection of quarters. A cleaning hysteria the likes of which I have never seen gripped the ship. Jan even painted his room as the fastest way to cover the dirt, but reverted to type by leaving the library full of his dirty glasses and his ashes on the deck. He now owes me $5.65 for borrowed tobacco. The ship bounces brightly on an easterly heading since noon.

April 17: Sunday morning again. Two jaegers follow easily and gracefully astern having no trouble matching our speed even against the wind. Our trouble against the wind is more apparent. We pitch so violently that in the middle of the night I got up to check the lab. The landlubbers in there are quite cavalier about securing things, leaving jars of fish pickled in formalin and microscopes freely teetering on shelves and benches. However, the ship must be built with a hinge; when I came stumbling from my quarters down to the boat deck, I found almost no motion in the lab aft on the main deck. Wild-eyed scrubbers are still blocking all passages, and swabs cannot be had for good money. No doubt a terrible fatigue will settle over all after 1000 hrs. We bounce and bounce and the creaking is curious on this still turbine ship; you never hear that over the engines on other ships—I thought it had gone out with wooden sailing vessels.

April 19: Reached my working area yesterday; remarkable how similar the relief is to the 22°N area, almost matched peak for peak so that I feel tempted to make extravagant topographic predictions. Here we will live and work (mostly work) for the next ten days. We set a buoy yesterday morning, a charming operation in cool breezes and pleasant sunshine. According to its flag its name is George's Big Bamboo, which was discovered when it was already out of reach, floating majestically like a wirework cherry tree constructed by Artzybasheff. We ride more easily on station and on westerly courses, bringing the sick back to life. I remain unbothered. Dusk brought a Homeric wine-dark sea, smoothly heaving without a ripple, with a pale yellow flash in the west, and a wideness and purity under the sky that filled the soul. Very quiet, too. At three in the morning up for a dredge. As time wears on, my dreams become more and more domestic, and being brusquely awakened at odd hours brings with it a sense of disorientation. Works wonders for the appreciation of home and relieves the desire for adventure. Reading Mark Twain, the art of precisely describing reality by overstating it. The dredge was successful, a peaceful slow job requiring skill and judgment exercised in no haste, and little brawn.

April 20: Completed our second east-west traverse at three in the morning, always such ungodly hours, and set the second buoy in darkness, a nice two-hour job at the end of which came sunrise in peculiar colors of green and yellow. Perfect weather and the sea as smooth as a pond. Afterward we succeeded in obtaining a 34-ft. piston core, the first really long one, and Geoff is now all smiles and happy, because he took the first failures rather personally. Working out the depth contours from the first two lines I find that the watches have been careless in marking time and depth and that there are many major errors. I shall have to read the records all over, a tedious job. Even so it is apparent that the two lines, only ten miles apart, are surprisingly different. We may have stumbled on another major fracture zone like the Vema to the north of us. The third line, run tonight, will, I hope, tell. As soon as we started on an easterly heading the wind picked up, and now we are again merrily pitching along. There is good fortune in this because it will delay our arrival till morning star time and provide the first full night of sleep in three days, just before we start on 72 hours of continuous coring. That will be rather trying, but since it is of the heart of my program I should not complain. Last night Alex heard the *Argo* on the radio, quite a feat since she is near Japan. He was quite pleased and had lots of news.

April 21: Woke up at 0300 worrying about the total absence of fixes since yesterday noon and could not go back to sleep. Getting up to the wheelhouse I found that not only do we not know where we are on the surface of the earth, but also the bottom profile is unaccountably different from the one just to the north that we ran yesterday. The black gloom on the bridge and this feeling of three-dimensional disorientation kept me up till morning star time but none showed. We did our best dead-reckoning, changed course, and arrived at George's Big Bamboo, our destination, within 15 minutes of the predicted time and with only a two-mile error. Sometimes things just go right, and we were overjoyed to see the iron cherry tree appear over the horizon in the emptiness.

Started on the long coring run with odd results. The ridge here has much less relief than farther north and appears, from our first five cores, to be entirely covered with sediment rather than bare like elsewhere. Could it be a lot older? I don't know what to make of it. Cores follow each other smoothly with two teams working, 12 hours on and 12 hours off; my task is to pick the station locations once every hour.

April 23: Yesterday morning, after losing three more corers and finding nothing but soft sediments and flat plains where the Mid-Atlantic Ridge ought to be, in the bright light of day we gave the whole area up for a bad job and turned north for the Vema fracture zone. Since I think there is some structural pattern to our depth data, this is not a hard decision because I am sure I can puzzle it out back home. So we went north and, for the first time, found the topography more or less as it ought to be, thus restoring the confidence of the watch in my judgment. Stopped for a deep dredge on an enormous 10,000-foot slope. It turned out to be just utterly long, hour after grinding hour of watching the wire go out and then come back in without so much as a bite for all seven hours. The Friday movie, however, was such a dud that I preferred to watch the meterwheel instead. For all our trouble we got a rock half an inch across, and at 0130 George and I decided to try once more by ourselves. The wire went over the side without trouble and, this being a fairly shallow haul, I went off to catch an hour's sleep while George stood the watch.

I immediately fell into one of those leaden dozes with formless dreams, a consequence of nine hours of sleep in three days, to wake at six when Vic shook my toes to tell me that we got hung up and lost half our main wire after three hours of vain struggle to come free. Usually I wake up with a snap, but this disaster took some time to sink in. This cuts the heart out of my program; not enough wire left to risk on dredging and just barely enough for some coring in shallower depth. I put on my usual disaster imperturbability while working out a program with emphasis on coring, hoping to make everybody feel better by this cheerful acceptance of the inevitable, but they remained gloomy as hell.

This prematurely terminates our third attempt to look closely at this area. Finding where we are seems unusually hard in this region because of the cloud cover and we have little luck in sampling, why I do not know. Something down there that God does not want me to see. It is aggravating but not uncommon and not as hard to live with as the general atmosphere of futility that seems to have infused this cruise from the beginning. Not only do we have more than our share of major mishaps, but even when things work, the results somehow are trivial. Even the transition from moderately sunny, not too cold, not too hot days to faintly starry nights is invariably accomplished through gray sunsets without fire and muggy evenings.

April 24: A minor problem with excessive drinking that I could not handle caused me to call the third mate in to quell the riot in the lab. It is amazing to observe the authority of even a lowly third mate, about the square of that of three chief scientists combined. A booby hitchhiked on the hydroboom, but when we turned east he did not like the course and went away.

April 25: The hitchhiker returned; this morning in the gray dawn he was sitting on the forward crosstree, nonchalantly dropping on the starboard bridge wing. Minutes later, the mate appeared in the crow's nest booing and shushing, apparently hit. No luck at all—the bird went on arranging his feathers and the mate disappeared in a cloud of unpleasant language, only to return with a broomstick. His silhouette against the paling sky stretched out on the crosstree and flailing the stick was unusual and compensated for a sleepless night. He scored a hit but the booby only removed himself aft, demolishing the anemometer by accident in his takeoff. The coring program is now over, only a partial success. As our equipment gets more sophisticated we can do much better, but also a higher premium is placed on knowing what is down there in order to use the right gear. This cruise I have not been a lucky guesser.

April 29: Suddenly we are in green water, so different from the translucent blue that has been around so long. We are still far from shore and this must be the Amazon mud plume making its way to the Caribbean. Rainy day with lots of squalls and fairly chilly so that my plans to finally acquire a suntan were foiled. Now I dream about going home and fret about how slow the days are moving. There is not nearly enough to do and people are hanging about. Even the supply of beer is gone.

May 1: Passed between Martinique and Dominica this morning, westward bound with flights of birds all around us. Martinique was smiling in the sunshine with brilliant green squares of cane fields on the darker mossy-looking rugged slopes above the steep cliffs falling into the ocean. Dominica over starboard was in a rain squall, vague, somber, menacing and extremely rugged—a beautiful sight. We set a northwesterly course for St. Thomas at noon and should be there tomorrow.

May 2: Day of arrival, very sunny, very choppy; the ship bumbles like an old bus on a Mexican secondary road, and it is clear that few have learned to control their walk. Antiroll tanks may cut down on the roll, but the resulting erratic short rolls and sharp pitches are difficult to manage. Soon there will be the sudden transition from ship to home across a gap that has no bridges. The abruptness and totality of this change always strikes me as one of the more peculiar features of my life. Looking back, however, the results of this month seem pretty worthwhile,* and there was much I shall be able to remember. Six months ashore and I may itch for another ride.

★ ★ ★

Now for the second cruise. This was a very different affair. It is clear that, in 1966, our two major frustrations had been a lack of means to find our location precisely enough at all times, and a failure, in the face of ever more precise work, to overcome the handicap of being separated from our objectives by more than 10,000 feet of water. The first of these was remedied in 1968 with the advent of navigation by satellites and the second was about to be cured on the FAMOUS (French-American Mid-Ocean Undersea Study) from which these fragments come.

FAMOUS Cruise, June–August 1974, Ponta Delgada, Azores, to Mid-Atlantic Ridge

June 19, 1974: Flying in low this morning over Ponta Delgada mole and saw *Marcel LeBihan* short-hauling the *Archimede,* and against the pier *Knorr* and *Lulu,* the latter looking disjointed like a disabled bridge span carried away in the flood of the century. Warm welcome on *Lulu,* strangely home. Notes I tacked on the bulkhead in February are still up, and my deck shoes, which I threw away in the Bahamas, somehow made it back to

* This turned out to be true, and a comprehensive and useful paper on the Vema fracture was published two years later.

my locker. After clearing a yard of miscellanea out of our quarters, just before Bob and Jim arrived and replaced it with another load, and making my bunk, I felt pretty good although tired after the pandemonium on the crowded flight from Boston.

June 20: Running a complex program by committee is always difficult, but if you do it with two ships tied up side-by-side where key people can hide, it becomes almost impossible. A day of meetings best summed up by Skip: "We heard that we would leave tomorrow and then maybe not; the other two hours were miscellanea." Slowly, though, essentials are taken care of.

June 22: Left port today after some confusion and general milling about. After all this effort and planning, the key base map of the Navy with all coordinates was discovered to be missing as the two ships were already separating, but in the nick of time it was handed over by long arms. Feels good to be outside—some chop and toward evening it became very chilly. The air is peculiarly vaporous, softening contours and creating ephemeral mists all the time. It rains frequently, although briefly, out of a clear sky. The test dive made it obvious that most people had forgotten their task, but the comedy acts put on by the pilots, perfectionists as they tend to be, indicated that all was well. My sea legs are not back yet. Back in port at midnight for further work.

June 23: Slow progress, back in port again after aborting the deep test dive because of weather. Yesterday was quiet and I used it to acquaint myself with all the equipment. There is a lot, and it is much easier to check it through on the beach than in the sub. Left at midnight, right into the weather. The *Lulu* does not ride well and bumped and shook in the strangest ways. At 0600 the wind was freshening, clouds low with rain and the sea stiff, with ten-foot swells and lots of blowing foam. So we canceled and went back to port. I slept—what else is there to do?

June 24: Sunday in port afforded one pleasure. It is apparently the local custom to drive onto the pier on Sunday afternoon and view the ships. The lines of cars parked on the concrete facing the ship in rows made us feel like a drive-in theater. The people are short and stocky, dressed always in dark clothes and, except for the smallest children, they never smile. Rain, mist, and a fresh breeze.

Driving light rain this morning as we left at 0800. Slept most of the day as did the others. The sea is not too bad, eight-foot swells quartering over the starboard bow with not too many whitecaps, a low, great, gray, scuttling sky with thin spraying rain, a primitive and joyless world populated only by us and a few shearwaters. Geysers and fountains rise through the grating on the foredeck as waves break between the catamaran hulls, and unless you like eating in wet pants, it takes skill to cross from quarters to galley hatch. Speed is 3.5 knots; hopefully that is temporary or it will take us five or six days to get there.

June 26: 130 miles in 36 hours and bumping all the way. The sun was out for a while this afternoon and we could be on deck but it has no warmth. Amazing how much sleep I can take—18 hours a day since Sunday. Otherwise wet, gray, sea state five, roll and bump and slam and hope that you can shoot into the galley hatch without getting drenched. Will be there Friday morning, same time as the French, but if the weather does not clear we will all just be steaming in circles.

June 27: Swells up to 12 feet and 35-knot wind and is it ever messy out. Generally, I fret about all this time lost without any accomplishment, a holdover from land where I scrimp and save on every hour. One little petrel joined us today, must be a lonely life.

June 29: Rough seas, two-knot speed, no place to be. And the rain did finally get to my bunk. A low point, I hope. The weather seems to be turning now; this morning on arrival at the site we found four ships cruising around and the *Glomar Challenger* just over the horizon. A real crowd after so many days on the big ocean. Big flurry to get ready and everybody happy, although still a little groggy after so much sleep. But at noon it blew up again and we canceled. Took some more instruction on the navigation system.

June 30: Well, first dive. We woke up to sunshine and good weather clouds, still a long swell and no wind. This will be a test dive—Bob and two pilots. Since Jim decided to do the navigation, I enjoyed a day of sunshine and anticipation on deck. Much oohing on return over all the gorgeous rocks and all this diddling with fascinating trivia that geologists are subject to, the stuff for papers with titles like, "Another Cone-Shaped Protrusion on a Marine Basalt." If I hadn't given up rock collecting years ago, none would ever have been labeled and packed. Curiously, on so extravagant a project, it seems that we forgot to load packing materials. The French were less lucky than we; *Archimede* still has problems, and *Cyana* was damaged in the launch and must return to port for repairs.

July 4: Three days no notes, this time for reasons of success rather than misery. I served my duty turn from navigator in the van, no chance for escape outside, to starboard observer in the sub, to dive leader yesterday. The weather has gotten nicer by the day and today it is millpond still, the wind indicator barely moves, and all that can be seen is a long low swell that does not really even move the *Lulu*. Monday was routine and the navigation system, already so fantastically sophisticated, got perked up a lot more. I still get lost in all the overlapping spheres of six ranges and a telemetry system, but we design more and more useful tricks and have *Alvin* pinned down to a few tens of feet almost all the time.

Tuesday was my first dive. Checked out the night before and things seemed comfortably familiar. Even so, it was a thrill to walk the plank and squeeze into the ball. Once inside, it is surprisingly snug and cozy and the tension falls away. Going down, the elaborate instruments checks make for a good transition. Finally, the pilot gets restless, the lights go on, something shadowy is down there, and we settle like a feather on a field of snow with protruding black boulders populated with lots of stalky, feathery things. The terrain was grandiose—large lava flows with a coarse, sneaking interwoven lace of huge tubes, often with collapsed tops and caverns underneath, steep fronts up to 40 feet high. I saw my first oceanic fault, six inches high, the next one two feet, but there they were. Then a bigger one, some 90 feet, looking like a pile of cookies that someone had sliced, or a gigantic wine rack with sliced basalt pillows for bottle bottoms. The last fault started with an innocent, very terrestrial looking little talus slope and then suddenly rose forever vertically into the black gloom [a thousand feet up, as we later found out]. I had to defend my porthole against Bob and his huge collection of cameras and recorders.

Then, upon return, we were told that the schedule required us to go back into port Friday. The weather is fine, the sub in shape, and we all want to go on diving, but the argument came to naught because important (not to us) people wish to get off or on in Ponta Delgada. The emphasis seems subtly changed from a diving program to something else, vague, big, and uncontrollable. I had a fit, but fortunately snapped out of it before we went to *Knorr* to talk it over. Another dive this morning; things are really smooth, finally.

July 5: Busy day on board with a bunch of French visitors. Things go right and they are impressed with the operation. We recovered at 1500 in worsening weather and after much chaos transferred to *Knorr* to go into port. A hairy thing with 10-foot seas bouncing the 12-foot whaler up and down *Knorr*'s immense side. The Fourth of July cake got over alright, however; we dropped the French off and now live in the oceanic Hilton Hotel, an alien ship with an alien crew and slightly vulgar after *Lulu*'s honest primitivity.

July 7: Back in Ponta Delgada with the usual crummy weather, drizzle, muggy, low overcast. The days on the way were spent on the data. It went a lot faster than we thought, but different team members have different aptitudes, and it is going to take a long time. *Knorr* reminds me of the *Glomar Challenger,* a big floating lab and residence hall, people everywhere and work going all the time in all nooks and crannies. It is stimulating in a sort of monastic way, and when you can't sleep you can always climb some ladders and join somebody at work.

July 11: At sea again. After several more days of data grind I finished

The Sun, Serene, Sinks into the Slumberous Sea by Ralph Albert Blakelock.

most of the computer work and discovered that I had not been off the ship yet for a breath of fresh air. So much for the charm of foreign ports. There was a crowd on deck, various people returning from shore leave with an amazing collection of ladies in tow, a lot of kissing, tears, waving, lines off, put *Lulu* in tow position, and, when just under way, several characters turned out to be on the wrong ship. When that was resolved, we moved out in stately fashion, whistles blowing, horns tooting, flags dipping, and the whole town cheering. Outside the ocean was smooth with a long, low swell rippled by a light northeaster; and the round green hills and numerous white houses, the occasional haciendas, the many large white churches, and the imposing black sea cliffs all were softened by the misty air that seems so typical of this island. Clouds form and disappear into nothingness and there is never anything sharp or sparkling. Everything is made lovely and mysterious by it, but we have yet to see the summit of the large volcano to our east. The sun set orange in the mist and I sat up forward for a long time contemplating the peace.

July 15: Woke up at six and treated myself to a shower before the dive. Beautiful calm day. All tasks are getting simpler and the hectic hassle seems over. The dive was quite short because we climbed up a vertical wall to get samples and used a lot of power, sometimes perched precariously on little ledges with the black abyss below. Jim and I appear very compatible in this precise approach, and the fault turned out to be a great deal more complicated than I had thought. Back at the surface early, we enjoyed a peaceful evening. As long as we can stay out of the channels of hierarchy everything is just fine, but alas, there is the radio ringing as often or more so than the phone in my office, and out of the sunset comes *Knorr,* no doubt with some upsetting news.

July 16: Made my first real mapping dive today connecting two previous explorations. Considering the short range of our vision, it is remarkable how well it all hung together. I am now much more able to place things in context in three dimensions, although it does require constant attention because you never have a good overview and, in the darkness, cannot make sketches easily. I miss perhaps a little the early excitement; it now seems such a normal thing to do, and down on the bottom it is a fairly routine environment. In fact, the sense of unreality is much stronger when we are back on *Lulu;* then the strange universe so brightly lit nearby and so quickly fading into gloomy cliffs beyond is a thing that haunts me in my dreams. Perhaps it ceases to exist when we are not there.

July 19: Yesterday in the evening back to *Knorr* to catch up on the data processing. It always seems to have to be me. *Knorr* is an odd ship. Since I came on board several complex and interesting operations have taken place, but virtually nobody is aware of them, lost as they are inside the ship. You cannot even really tell if *Knorr* is stopped or not. Several co-workers had been unaware of my five-day diving absence. This business of continuously switching ships in mid-ocean is, to say the least, confusing.

July 22: Well, life at sea is not always dull. Barely had I finished writing, had dinner, and settled down to watch a movie that was a good deal sexier than I need after six weeks at sea, when the ship's phone started ringing (this place has more phones and radios and calls, day and night, than even my office). It appeared that *Alvin* had developed serious trouble and had to go back to port. Rather than breaking *Knorr*'s routine, apparently important, *Lulu* was going to make it on her own in five or six days rather than two in tow. I lost the argument handsomely and find that difficult to take, on top of the waste of time. I guess I am no longer very good at following and am beginning to think that I should have stayed clear of the whole project. And yet, the seafloor is so beautiful, so out-of-this-world beautiful. The misty, grandiose, mysterious landscape of craggy black and snowy, pillowy white set in the foreground with brilliantly lit small, sharp, and perfect vignettes of rock, coral, and sponge, is unforgettable and I cannot do justice to it. I wish I could draw, it would be so superior to photography. It is an exceptional privilege to have seen this. Already, thinking that my last dive might just have been behind me, there is a sharp pang of nostalgia.

July 21: We must be in port again. The ship doesn't roll, delicious but fattening Portuguese pastries have shown up in the mess, and the coming and going of poorly identified people has taken on crisis proportions. Most of the hullabaloo is about the big press conference and the massive arrival of French and U.S. dignitaries, something that the dive team is not really looking forward to.

August 5: Finished what may have been my last dive and transferred, once again, to *Knorr.* The weather is gorgeous and the dive covered the central volcanoes which I hadn't seen before. Tremendous lava flows, far more impressive, jagged, and bizarre than anything I had seen before. The landscape was magnificent and very pure and pristine, not fouled by anything since God made it yesterday, and so different from the tired and worn scenery of the fracture zone dives. This pleases the soul: it is crisp, strong, and almost brutal.

August 8: Alvin is sick with poor batteries. It was long seen coming but now it is serious enough to cancel a dive and raise doubts about the remaining ones. Very complicated three-way, two-language radio conversations follow which suffer from bad atmospherics and the disappearance of essential individuals at critical moments. It took several hours, accomplished little, and was hilariously funny.

August 9: Well, the die was cast yesterday, late in a hot and glassy calm afternoon. *Alvin* must return to port to drop the batteries and the whole strange adventure is unexpectedly over. We pulled up all the transponders, put three protesting filmmakers on *Lulu,* transferred the diving team to *Knorr,* hitched up the tow, and were on the way, suddenly in an atmosphere of plane reservations, thoughts of home, and abrupt dissolution of the tight bonds in this community of two months and of three years' preparation. Overnight, we became again the loosely knit conglomerate of randomly picked people that we really are. I sat last night for a long time watching it get darker in yet another beautifully still evening, watching several whales go by, as they had all day—a surprising display of what one considers as an almost extinct animal—and pondering adjustment. Inevitably, one of these nights, one of these years, it will really turn out to be the last one of my life, and I shall not even know it, I suppose.

August 10: Total disintegration now and a good deal of weepy farewells. The souvenirs are fanciful but not very practical. We are loading hundreds of sheet-metal milk cans, while *Lulu* went in for ancient wooden wine barrels. A craze in brass cowbells was set off, and there is enough basketwork to contain Cape Cod's laundry. Most seem to go for bulk rather than refinement, but then, they do not have to fly it home.

And now the transition to reality at four tomorrow morning. Or is it?

Battle scene from the comic operatic fantasy "The Seafarer" by Paul Klee.

"They knew the rhythm of the tides by memory, the significance of the water's coloring, the location of the reefs. They used to say that they could read the form and wrinkles of the water, but the truth is that in their knowledge of the sea there was habit, the five senses, and perhaps one more."

Jose Maria Gironella, *Fantoms and Fugitives,* 1964

The Return of Neptune by John Singleton Copley.

READER'S GUIDE

In this book, I have woven the story of ocean science in a somewhat unconventional manner. It is not so much that I have used exotic threads, although an occasional odd one is in evidence, but that the pattern is different from the usual orderly treatment of the subject. This results mainly from my insistence on a historical theme, in all senses of the word: history of the earth, history of life, history of man, history of oceanography, history of one oceanographer. I must attempt to restore the balance here by providing access to additional reading.

This is more easily said than done. I envy the humanist and social scientist because, impenetrable as their style or jargon may be at times, their subjects generally are not. Moreover, they write books, and books can easily be found. Physical scientists, on the other hand, produce writings that are, jargon or not, moderately hard to penetrate, and they prefer to present their results in small pieces—articles mostly—in exotic forbidding places like the *Journal of Geophysical Research.* Thus, my choice is limited—and even more so because for one reason or another ocean scientists have tended to neglect the general audience even more than have their colleagues in other disciplines.

A curious reader now may feel the desire to explore the field in its breadth in a more conventional manner. This commendable wish leads inevitably to some college textbook in oceanography. There are many of those, but in this context I particularly like the following:

Ross, David A. *Introduction to Oceanography.* 2nd ed. New York: Appleton-Century-Crofts, 1976.

Weyl, Peter K. *Oceanography: An Introduction to the Marine Environment.* New York: John Wiley & Sons, 1970.

More geologically oriented, but short and complete, is:

Turekian, K.K. *Oceans.* Foundations of Earth Science Series, 2nd ed. Englewood Cliffs, New Jersey: Prentice-Hall, 1976.

Books on the history of oceanography are as yet very few, but they provide a beautiful insight into the evolution of the science and the scientists and also, if well done, teach concepts and facts in a context that is particularly pleasant to absorb. The best one I know is:

Schlee, Susan. *The Edge of an Unfamiliar World: A History of Oceanography.* New York: E.P. Dutton & Co., 1973.

Somewhat less readable but full of unusual anecdotes and historical oddities, although it is restricted mainly to European history and runs only to the end of the 19th century:

Deacon, Margaret. *Scientists and the Sea, 1650-1900: A Study of Marine Science.* New York: Academic Press, 1971.

During the last two decades, ocean politics has become a new and absorbing field and understanding it is to some extent a necessary part of understanding ocean science. The best (and only) guide, written in a somewhat complex manner, with a strong flavor of Washington is:

Wenk, Edward, Jr. *The Politics of the Ocean.* Seattle: The University of Washington Press, 1972. (The author was Executive Director of the National Marine Council and an observant witness of much of the substance and foolishness of the last 15 years.)

And so we proceed to the geological revolution. Naturally, a great deal has been said about the subject, but in my opinion one book stands out above all others—for the purpose of this reader's guide. Walter Sullivan is a senior science editor of the *New York Times* and for many years has not only followed the scientific scene across the world but has often been there at the right moment and with the right people. Couple this with an unusually penetrating mind, a charming style, and a very thorough knowledge and erudition and you have a combined historical account and treatment of the principles of continental drift that is hard to match.

Sullivan, Walter. *Continents In Motion: The New Earth Debate.* New York: McGraw-Hill, 1974.

A bit more technical, because it consists of articles reprinted from *Scientific American,* but an elegant overview of most of the aspects we have discussed in Chapters Two through Four (with the exception of marine biology), is:

Continents Adrift and Continents Aground: Readings from the Scientific American. Edited by J. Tuzo Wilson. San Francisco: W.H. Freeman & Co., 1976.

Rather light on the science but beautifully illustrated in the traditional *National Geographic* manner and very worthwhile is:

Matthews, Samuel W. "The Changing Earth." *National Geographic,* vol. 143, 1973, pp. 1-37.

Finally, a small pocketbook, a bit dry looking but easy to read, also deals with this subject in a comprehensive way:

Tarling, Don and Maureen. *Continental Drift: A Study of the Earth's Moving Surface.* New York: Anchor Books, Anchor Press, 1975.

As to marine biology, fisheries, and other such subjects—at least as far as they are relevant to the content of this book—there is not very much from which to select. Of course, there are endless numbers of beautiful books on life on shores and coral reefs and such things, but these subjects fall outside my scope. The one that I personally rather like, because it is concise, easy to read, and reasonable in attitude, is:

Bardach, John. *Harvest of the Sea.* New York: Harper and Row, 1968.

We are equally short of references on climates and paleoclimates. The source of much of the information in Chapter Four is an appendix to a publication of the National Academy of Sciences, but it does not make for light reading.

Understanding Climatic Change. United States Committee for the Global Atmospheric Research Program. Washington, D.C.: National Academy of Sciences, 1975.

Utterly readable, full of fascinating details, a bit chaotic as such historical books sometimes appear, and not completely in tune with my own thinking is the following book. Nobody interested in climates should miss it.

Ladurie, Emmanuel Le Roy. *Times of Feast, Times of Famine: A History of Climate Since the Year 1000.* New York: Doubleday, 1971.

As the public's concern about weather aberrations and the future course of climate has increased, authors have taken advantage of the potential market. One of these is the following, whose discussion of climate and the world food situation is of interest:

Schneider, Stephen H., and Lynn E. Mesirow. *The Genesis Strategy: Climate and Global Survival.* New York: Plenum Publications, 1976.

Much more concise, but very much better, is a short article in the magazine *Science.* It represents but one of the many possible views but I am personally prejudiced in its favor.

Bryson, Reid A. "A Perspective on Climatic Change." *Science,* vol. 184, 1974, pp. 753-60.

Except for what has been cited in the text, I am unfortunately unable to suggest further reading on the subject of salt and sea level. Happily, the situation is better for the subject of marine mineral resources. There is a plethora of federal government publications, National Academy reports, and other such things, but the following books will sharpen the picture beyond my very condensed summary. The first is perhaps already a little out of date and the author is an incurable optimist, but it does contain a lot of information, much of it still good. The second is newer and more general since it deals with all of man's interactions with the sea, but it springs from a pair of sober minds, and it has a limited, but well chosen, list of further readings.

Mero, John L. *The Mineral Resources of the Sea.* New York: Elsevier, 1965.

Skinner, Brian J., and Karl K. Turekian. *Man and the Ocean.* Foundation of the Earth Sciences Series. Englewood Cliffs, New Jersey: Prentice-Hall, 1973.

A good historical account of the evolution of concepts of freedom of the sea and territorial limits can be found in:

Alexander, Lewis M. *Offshore Geography of Northwestern Europe: The Political and Economic Problems of Delimitation and Control.* Third Monograph of the Association of American Geographers. Chicago: Rand McNally, 1963.

Concerning the recent fray over the law of the sea, there is not yet much literature, but what there is can generally be found in the *Ocean Development and International Law Journal.*

Finally, anyone interested in the oceans should once in a while sit down and ponder a map at his leisure, in particular a map showing the topography of the ocean floor. Apart from a small globe issued by the National Geographic Society, by far the most beautiful map for this purpose is:

Heezen, Bruce C., and Marie Tharp. *The Floor of the Oceans.* New York: American Geographical Society, 1975.

INDEX

INDEX

ABOUT THE AUTHOR

Tjeerd H. van Andel, renowned marine geologist and a member of the first scientific expedition to view and map the Mid-Atlantic Ridge from a research submersible, has devoted his professional life to the investigation of the undersea world. He has been involved in tectonic ocean mapping, deep-sea drilling, ocean exploration, mineral resource assessment, manned undersea activities, and paleo-oceanography. His numerous scientific papers contain pioneer contributions to his specialty, earning him international recognition including membership in the Royal Netherlands Academy of Sciences.

Professor of geology at Stanford since 1976, Dr. van Andel was born in Rotterdam and raised in Indonesia. He received his bachelor's, master's, and doctor's degrees from the University of Gronigen in the Netherlands. After teaching at the University of Wagenigen, Netherlands, for two years, he turned to industry, spending the next six years as a senior scientist with Shell Oil Company, first in Amsterdam and then in Maracaibo, Venezuela. He resumed his teaching career when he came to the United States in 1957, accepting an academic appointment at the University of California's Scripps Institution of Oceanography in La Jolla. In 1968 he joined the faculty of Oregon State University at Corvallis where he became the head of its School of Oceanography's Marine Geosciences Division, developing it into national prominence before accepting his appointment in the School of Earth Sciences at Stanford University.

Literally "driven out to sea" by his native curiosity as well as by his scholarly interests, Tjeerd van Andel is completely at home two miles underwater. He was a participant in the 1974 international expedition termed "FAMOUS" which explored the divergent plate edge on the Mid-Atlantic Ridge. The U.S. research submersible *Alvin* took van Andel and his colleagues to depths of over 12,000 feet for the historic seafloor mapping which became the subject of wide media coverage including two articles in the *National Geographic.* A second *Alvin* exploration in February 1977 took van Andel to greater depths for study of the fumaroles and fissures exuding hot lava from the seafloor on the Galapagos Rift. This privileged research of the undersea world will continue to enrich Dr. van Andel's scientific knowledge and contribution to society.